CONDUIT BENDING

2020 EDITION

Dr. Alan W. Stanfield

A Note from the Editor...

Ugly's Conduit Bending is a quick, on-the-job reference specifically designed to provide the most commonly required information on how to properly bend conduit, including information on bending types and techniques. An ideal tool for electricians, contractors, instructors, and students, this essential pocket guide uses diagrams, calculations, illustrations, photos, and quick explanations to ensure bending is completed safely and correctly.

We salute the National Fire Protection Association for its dedication to the protection of lives and property from fire and electrical hazards through the sponsorship of the *National Electrical Code®*.

National Electrical Code® and *NEC®* are registered trademarks of the National Fire Protection Association, Inc., Quincy, MA.

JONES & BARTLETT
L E A R N I N G

TABLE OF CONTENTS

Acknowledgments

The author would like to thank the following individuals for their contributions to this edition:

Rebekah Hosch

Sean Ivey

Deanna Phillips

Shelley Worley

⌨ INTRODUCTION

In the electrical industry, the ability to perform efficient and accurate conduit bending can be a very valuable asset. Quality conduit bending requires both mechanical skill and mathematical ability to perform safe and effective installations.

The purpose of conduit bending is to provide a safe and effective electrical installation using techniques that install customized bends in conduit. This approach reduces the number of prefabricated fittings and joints required. Bending conduit is a more efficient and economical method of installing conduit than installing prefabricated fittings such as corner couplings and prefabricated offsets.

This *Ugly's* guide provides the formulas, tables, and step-by-step instructions needed to make common conduit bends. The photographs included demonstrate proper bending technique. Keep in mind, however, that effective conduit bending is a skill that must be learned through extensive practice. Use this manual to practice and ultimately master the art of conduit bending.

CONDUIT-BENDING SAFETY

When working in any electrical environment, safety is always the primary consideration. Often, safety is neglected due to time constraints, schedule demands, or simply a lack of concern. No one should be so confident in his or her abilities that safety is compromised.

Conduit plays an important role in the electrical industry by providing support for conductors as well as protection of the conductors and the electrical system. Although conduit runs are often installed in locations that are covered by drywall or concealed in block walls, there are many times when conduit is exposed. Not only is it important to make conduit bends with aesthetics and maintenance of the cross-sectional area in mind, but one must also consider the safety of the installation process and job site.

Conduit bending involves a number of activities that can cause minor to severe injuries. Many of these injuries can be avoided, however, by taking simple safety precautions. Safe conduit cutting, bending, threading, and installation begin with selecting the appropriate tools and personal protective equipment (**Table 1**).

Table 1   Conduit-Bending Tool/Equipment Checklist

Proper Attire (Figure 1)
Long pants made from a heavy, durable material
Properly fitting shirt, tucked in
Sturdy work boots—ideally, steel-toed
Long hair secured
Jewelry removed

Personal Protective Equipment
Eye protection (select one):   • Safety glasses—offer front and limited side protection   • Safety goggles—protect eyes from debris or dust that can enter the eye at any angle
Helmet (hard hat)
Gloves

CONDUIT-BENDING SAFETY

Cutting conduit (select one):
- Common hacksaw—used for electrical metallic tubing (EMT) and small conduit
- Pipe cutters—used for rigid pipe that will be threaded
- Polyvinyl chloride (PVC) cutters

Conduit reamer—removes burrs from cut conduit

Torpedo level

Creating knockout holes in cabinets and junction boxes (select one):
- Step bits
- Knockout punches

Pulling wire through conduit (select one):
- Fish tape
- Cable tugger
- Winch

Wire-pulling lubricant

Figure 1 Appropriate conduit-bending attire and personal protective equipment.

CONDUIT-BENDING SAFETY

Safe Work Space

Conduit bends need to be made on flat and level surfaces. Bends should not be made on uneven surfaces, and it is not a good idea to make bends on surfaces that are unstable, such as sandy soil. The area also needs to be clear of clutter and debris. Ensure that you have adequate space to move the conduit into different positions without striking other objects or people.

Ladder Safety

Falling from a ladder at any height can lead to serious injury or even death. Ladder safety is especially important when working with electricity. Always use a fiberglass ladder, which will restrict the flow of electricity if you accidentally come in contact with electrical conductors. Aluminum or other metal ladders should never be used while performing electrical tasks, even if you are certain that the power has been removed from the circuit. Likewise, wooden ladders should be avoided because electrical currents can flow through damp or wet wood.

When using a ladder, always adhere to the following general safety precautions:

1. The base of an extension ladder should be positioned 1 ft (0.30 m) away from the top of the ladder for every 4 ft (1.22 m) of the ladder height. The minimum height that an extension ladder should extend past the top of the structure is 3 ft (0.91 m) (**Figure 2**).

2. Ensure that the surface on which the ladder is placed is dry and stable to avoid slipping or shifting. If possible, have someone support the bottom of the ladder while it is in use. When using an extension ladder on a wooden surface, consider nailing a 2- × 4-in. piece of lumber at the toe of the ladder to provide a slip guard.

3. When using a step ladder, securely lock the supports and place the ladder on the most level surface available. Carefully read the manufacturer's warning label for the proper weight requirements and any other warnings provided.

Figure 2 Safe ladder placement.

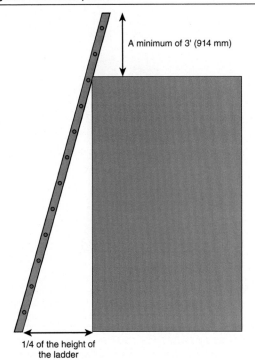

A minimum of 3' (914 mm)

1/4 of the height of
the ladder

Tool/Equipment Damage

Two of the most common on-the-job safety violations are misuse of tools and use of damaged tools. Always use the correct tool or piece of equipment for the job you are performing. For example, you should not use a screwdriver as a chisel or lineman pliers as a hammer (**Figure 3**). Tools are designed with specific purposes in mind. When they are used for purposes that are not intended, accidents may occur.

CONDUIT-BENDING SAFETY

Figure 3 A pair of lineman pliers that has been misused as a hammer.

Courtesy of Alan W. Stanfield.

If a tool or piece of equipment has been damaged, you should repair it or find an undamaged replacement. Tools or pieces of equipment should be replaced if they can no longer be used safely. Cutting and crimping tools, such as lineman pliers, are frequently used and are often damaged in the field. For safety reasons, pliers and other hand tools should be replaced after being damaged. If a damaged screwdriver continues to be used in the field, for example, it will cause further damage when used to tighten or remove screws.

Lockout–Tagout

Following correct lockout–tagout procedures (see page 147) can be a question of life or death. Although locking and tagging equipment may require a little more time on the job, it is time well spent.

6

Figure 4   Hasp devices.

Courtesy of Ideal Industries, Inc.

Many different lockout–tagout devices are available, but all serve the same basic function: restricting unauthorized personnel from powering circuits or pieces of equipment that are being repaired or replaced. Hasp devices, for example, accommodate padlocks to ensure safety (**Figure 4**). Along with tags, they are placed on the electrical disconnect that controls the equipment being repaired.

Each worker attaches his or her own lock to a device, and that lock can be removed only by the assigned individual. Tags applied to the lockout device indicate the individual who is performing labor on a piece of equipment and other warning information, such as the voltage and the date of repair.

CONDUIT-BENDING SAFETY

A lockout–tagout center is often located near the equipment that requires servicing. These centers are usually prominently marked and provide locks, tags, and any other equipment needed to ensure electrical safety (**Figure 5**).

Figure 5 A typical lockout–tagout station.

Courtesy of Ideal Industries, Inc.

🔌 CONDUIT

Electrical conduit is available in a wide variety of sizes, compositions, and types (**Figure 6**). It can be purchased in both metallic and nonmetallic forms. Metallic conduit is generally sold in lengths of 10 ft (3.05 m). Flexible conduit is sold in rolls.

Figure 6 Rigid metal conduit (RMC) alongside a piece of ¾-in. and ½-in. thin-wall EMT.

Note: EMT is categorized as tubing rather than conduit, but it is used as conduit in wiring installations.

🔌 CONDUIT

The type and design of electrical conduit used on a job site depend on a number of factors. The *National Electrical Code*® (*NEC*®) outlines guidelines for the installation and proper usage of all available types of conduit:

Article 342. Intermediate Metal Conduit: Type IMC **(Table 2)**

Article 344. Rigid Metal Conduit: Type RMC **(Table 3)**

Article 352. Rigid Polyvinyl Chloride Conduit: Type PVC **(Table 4)**

Article 353. High-Density Polyethylene Conduit: Type HDPE (Table 4)

Article 358. Electrical Metallic Tubing: Type EMT **(Table 5)**

Article 362. Electrical Nonmetallic Tubing: Type ENT **(Table 6)**

NEC® Chapter 9, Table 4 provides dimension and percent area information related to each of these types of conduit. Excerpts from *NEC*® Chapter 9, Table 4 provide common trade sizes and conduit types in the following tables.

Conduit Dimensions

Table 2 Dimensions and Percent Area of IMC

Trade Size	60% Fill, for Nipples		40% Fill, for More Than Two Wires	
	in.²	mm²	in.²	mm²
½	0.205	133	0.137	89
¾	0.352	226	0.235	151
1	0.575	372	0.384	248
1¼	0.988	638	0.659	425
1½	1.335	859	0.890	573
2	2.178	1405	1.452	937
2½	3.081	1985	2.054	1323
3	4.753	3069	3.169	2046
3½	6.351	4093	4.234	2729
4	8.179	5235	5.452	3490

Modified from *NEC*® Chapter 9, Table 4, Article 342.

CONDUIT

Table 3   Dimensions and Percent Area of RMC

Trade Size	60% Fill, for Nipples		40% Fill, for More Than Two Wires	
	in.²	mm²	in.²	mm²
½	0.188	122	0.125	81
¾	0.329	212	0.220	141
1	0.532	344	0.355	229
1¼	0.916	591	0.610	394
1½	1.243	800	0.829	533
2	2.045	1319	1.363	879
2½	2.919	1882	1.946	1255
3	4.499	2904	3.000	1936
3½	6.006	3877	4.004	2584
4	7.729	4990	5.153	3326
5	12.127	7830	8.085	5220
6	17.495	11292	11.663	7528

Modified from *NEC*® Chapter 9, Table 4, Article 344.

CONDUIT

Table 4 Dimensions and Percent Area of PVC, Schedule 40, and HDPE

Trade Size	60% Fill, for Nipples		40% Fill, for More Than Two Wires	
	in.²	mm²	in.²	mm²
½	0.171	110	0.114	74
¾	0.305	196	0.203	131
1	0.499	321	0.333	214
1¼	0.872	561	0.581	374
1½	1.191	769	0.794	513
2	1.975	1274	1.316	849
2½	2.817	1817	1.878	1212
3	4.361	2816	2.907	1877
3½	5.842	3766	3.895	2511
4	7.532	4855	5.022	3237
5	11.856	7649	7.904	5099
6	17.140	11060	11.427	7373

Modified from *NEC*® Chapter 9, Table 4, Article 352 and 353.

Table 5 Dimensions and Percent Area of EMT

Trade Size	60% Fill, for Nipples		40% Fill, for More Than Two Wires	
	in.²	mm²	in.²	mm²
½	0.182	118	0.122	78
¾	0.320	206	0.213	137
1	0.519	333	0.346	222
1¼	0.897	581	0.598	387
1½	1.221	788	0.814	526
2	2.013	1299	1.342	866
2½	3.515	2270	2.343	1513
3	5.307	3421	3.538	2280
3½	6.927	4471	4.618	2980
4	8.852	5712	5.901	3808

Modified from *NEC*® Chapter 9, Table 4, Article 358.

CONDUIT

Table 6 Dimensions and Percent Area of ENT

| Trade Size | 60% Fill, for Nipples | | 40% Fill, for More Than Two Wires | |
	in.^{2}	mm^2	in.^{2}	mm^2
½	0.171	110	0.114	73
¾	0.305	197	0.203	131
1	0.499	322	0.333	215
1¼	0.872	562	0.581	375
1½	1.191	769	0.794	512
2	1.975	1274	1.316	849

Modified from *NEC*® Chapter 9, Table 4, Article 362.

Manufacturer Markings

Most conduit will have some type of manufacturer marking indicating the size. Some EMT conduit, for example, will have markings that indicate every 12 in. (305 mm) of length. This system allows for quicker cutting and bending of conduit in the field. These markings are more commonly found on EMT and PVC conduit than on rigid conduit.

Rigid conduit size is usually indicated by the color of the thread protector cap rather than by a size stamped directly on the conduit (**Table 7**). For RMC, the following color scheme is used:

- Whole-inch sizes of conduit have blue thread protector caps.
- Half-inch sizes have black thread protector caps.
- Quarter-inch sizes have red thread protector caps.

IMC also has color-coded thread protector caps:

- Whole-inch sizes have orange thread protector caps.
- Half-inch sizes have yellow thread protector caps.
- Quarter-inch sizes have green thread protector caps.

CONDUIT

Table 7 Thread Protector Cap Colors for RMC and IMC

Trade Size	Thread Protector Cap Color	
	RMC	IMC
½	Black	Yellow
¾	Red	Green
1	Blue	Orange
1¼	Red	Green
1½	Black	Yellow
2	Blue	Orange
2½	Black	Yellow
3	Blue	Orange
3½	Black	Yellow
4	Blue	Orange
5	Blue	Orange
6	Blue	Orange

Data from Steel Tube Institute of North America Guidelines for Installing Steel
Conduit/Tubing [www.steeltubeinstitute.org]. Accessed May 18, 2010.

Conduit Fittings and Accessories

Many different types of conduit fittings, couplings, straps, and other
accessories are associated with installing conduit systems (**Figure 7**).
Such fittings include the following items:

- **Box connectors.** Used to terminate or connect conduit to a box
  or other type of enclosure. Some fittings are die cast; others are
  made from steel.
- **Conduit couplings.** Used to connect two pieces of conduit.
  Include screw type, compression type, and other fittings for
  various functions (e.g., 90-degree elbow with a removable cover
  and a premade box offset).
- **Conduit straps.** Used to secure conduit to walls.
- **Conduit clamps.** Used to connect conduit to a strut or channel
  material.

CONDUIT

Figure 7 Conduit fittings and accessories.

A. Box Connector

B. Screw-Type Fitting

C. Compression-Type Fitting

D. 90-Degree Elbow with Removable Cover and Premade Box Offset

E. Conduit Straps

Courtesy of Alan W. Stanfield.

F. Conduit Clamps

CONDUIT

Proper burial depths are required by the *NEC®* Table 300.5. An excerpt from *NEC®* Table 300.5 is provided for reference **(Table 8)**.

Table 8 Minimum Cover Requirements, 0 to 1000 Volts, Nominal, Burial

Location of Wiring Method or Circuit	RMC or IMC		Nonmetallic Raceways Listed for Direct Burial Without Concrete Encasement or Other Approved Raceways	
	in.	mm	in.	mm
All locations not specified below	6	150	18	450
Under a building	0	0	0	0
Under minimum of 4-in. (102-mm)-thick concrete exterior slab with no vehicular traffic and the slab extending not less than 6 in. (152 mm) beyond the underground installation	4	100	4	100
In trench below 2-in. (51-mm)-thick concrete or equivalent	6	150	12	300
One- and two-family dwelling driveways and outdoor parking areas, and used only for dwelling-related purposes	18	450	18	450
In or under airport runways, including adjacent areas where trespassing is prohibited	18	450	18	450
Under streets, highways, roads, alleys, driveways, and parking lots	24	600	24	600

Note: This table represents an excerpt from Table 300.5 of the 2020 *NEC®*. For more complete information, consult the *NEC®*.

Modified from *NEC®* Table 300.5.

CONDUIT BENDERS

Four basic types of benders are available to the electrician: hand, mechanical, electric, and hydraulic. **Table 9** outlines the key considerations when selecting the most appropriate bender for your bending needs. The type and style of benders will vary based on field conditions, price, the size of the conduit being bent, and the volume of conduit bends needed.

Table 9   Recommended Bending Techniques for Different Types of Conduit

Type of Conduit	Recommended Bending Techniques
IMC	Electric bender, mechanical bender, hydraulic bender, hickey bender, or hand bender (for smaller sizes of IMC)
RMC	Electric bender, mechanical bender, hydraulic bender, or hickey bender (trade sizes ½ in. and ¾ in.)
PVC	Hot box or heat blanket
HDPE	Hot box or heat blanket
EMT	Electric bender, mechanical bender, or hand bender
ENT	Hot box or heat blanket

CONDUIT BENDERS

Hand Benders

Hand benders (**Figure 8**) are good choices for job sites where access to electrical power is unavailable and in locations where mobility is needed. They are beneficial to jobs requiring a large amount of bending to be performed because the technique involved here is the least expensive method of bending conduit. The most common sizes of EMT that are bent using hand benders are ½ in. and ¾ in. In addition, 1-in. EMT benders are available.

Figure 8   A hand bender.

Courtesy of Ideal Industries, Inc.

CONDUIT BENDERS

Hand benders are difficult to use on conduit that is larger than ¾ in. Although 1-in. hand benders can be used in the field, most electricians prefer to use a mechanical or electric bender for conduit larger than ¾ in. The 1-in. hand benders are bulky and require much more effort to use than ¾-in. benders **(Figure 9)**.

Figure 9 Hand bender marks.

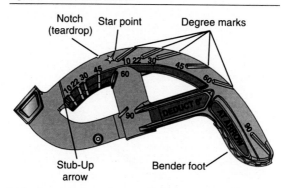

Modified with permission of Greenlee Textron, Inc.

Hand Bender Marks

- **Stub-up arrow:** the arrow closest to the end of the bender head. This mark on the bender is used to make stub-up, saddle, offset, and other bends.

- **Star point:** marks the back of the 90-degree bend or the center of the bend radius. This mark on the bender is used to make back-to-back bends.

■ **Notches** (or **teardrop**): indicates the back of a 45-degree bend, which is used on many three-bend saddles. The teardrop or notches in the bender can be found between the stub-up arrow and the star point. Some benders will have a teardrop, whereas other benders may have notches cut into the bottom of the bender head.

Offset Bend Tool

The offset bend tool is used to make small offset bends, such as the ones that enter boxes. Inserting the conduit and pressing the handle will create both bends required for an offset bend in one motion (**Figure 10**).

Figure 10 An offset bend tool.

Courtesy of Greenlee Textron, Inc.

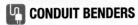

CONDUIT BENDERS

Hickey Bender

The hickey bender is used with rigid conduit and IMC. It is designed to be used like a multiple-shot hydraulic bender, which makes a number of small bends to form a larger bend. For example, this tool works by producing nine 10-degree bends to form a complete 90-degree bend. The hickey bender should not be used with EMT or thin-wall conduit because it is designed to support only rigid conduit and is used as a segment bender (**Figure 11**). The hickey bender is also useful for making tighter-radius bends and is available in a variety of sizes.

Figure 11   A hickey bender.

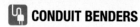

CONDUIT BENDERS

Mechanical Benders

Mechanical benders (**Figure 12**), which are also known as ratcheting or Chicago benders, are less portable than hand benders. Although they are more expensive, they can create more bends in a given amount of time than hand benders.

Figure 12 A mechanical bender.

Courtesy of Greenlee Textron, Inc.

🔌 CONDUIT BENDERS

The biggest advantages of this type of bender relate to its ability to bend large, rigid conduit while allowing for greater mobility than an electric or hydraulic bender. These devices do not require electrical power. In addition, they have an advantage over hand benders because the use of a mechanical bender requires less physical labor. The ratcheting action of the mechanical bender reduces the amount of force needed to bend the conduit.

Mechanical Bender Marks

The markings on a mechanical bender generally include a bending degree indicator to assist the operator in determining the appropriate bending angle (**Figure 13**). The bender usually includes a series of tables and diagrams showing how to make stub-up bends, offset bends, and saddles (**Figure 14**). Always make sure the angle dial is calibrated before making any bends.

Figure 13 Parts of a mechanical bender.

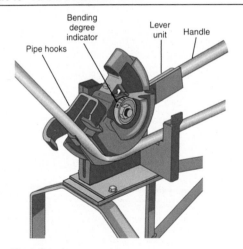

Courtesy of Greenlee Textron, Inc.

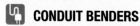

CONDUIT BENDERS

Figure 14   Operator reference materials.

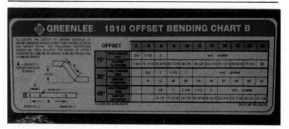

Courtesy of Alan W. Stanfield.

Electric Benders

Electric benders (**Figure 15**) are very useful on jobs that require precision bending and/or a high volume of bends. These devices are generally used to bend larger sizes of conduit, but most can be adjusted to accommodate a variety of sizes, often ranging from ½ in. to 4 in.

CONDUIT BENDERS

Figure 15 An electric bender.

Courtesy of Greenlee Textron, Inc.

Due to its size and weight, an electric bender will be most effective when it can be set up to bend large amounts of conduit in one static location.

25

🔌 CONDUIT BENDERS

Electric Bender Marks

The markings on an electric bender are similar to those found on a mechanical bender. Tables and other reference materials are located on the bender so that the operator has quick and easy access to the necessary information (**Figure 16**). Anyone operating the electric bender should be able to interpret the information concerning the basic operation of the machine, so as to make accurate bends.

Figure 16   Parts of an electrical bender.

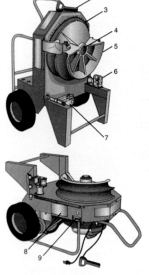

Parts List:
1. Handle
2. Pendant
3. Protractor
4. Main shaft
5. Bending shoe
6. 1½-in. to 2-in. roller support
7. ½-in. to 1¼-in. roller support (in storage position)
8. Bending instructions decal
9. Circuit breaker (on/off switch)

Courtesy of Greenlee Textron, Inc.

CONDUIT BENDERS

Hydraulic Benders

Hydraulic benders are often lighter than electric benders but offer similar strength. They are typically more mobile, partly due to their ability to be disassembled. These devices can be electrically operated or manually pump operated, which makes them versatile.

Hydraulic benders share some of the same limitations as mechanical and electric benders. Notably, they are powerful but can be expensive and difficult to navigate around the job site. Because hydraulic benders are easier to move and lighter than electric benders, they are often a better option if bending of large conduit is required.

Hydraulic benders are available in two general types: the one-shot bender (**Figure 17**) and the multiple-shot bender (**Figure 18**). The one-shot bender can complete a desired bend in one complete motion of the bender. The multiple-shot bender makes a series of smaller bends to form the overall complete desired angle. This process takes more time because the conduit must be repositioned after each smaller segment bend is created.

Figure 17   One-shot hydraulic bender.

Courtesy of Greenlee Textron, Inc.

CONDUIT BENDERS

Figure 18 Multiple-shot hydraulic bender.

Courtesy of Greenlee Textron, Inc.

Hydraulic Bender Marks

Markings on hydraulic benders vary. Most hydraulic benders provide reference tables or charts that allow the operator to determine how to make the proper bend.

![icon] CONDUIT BENDERS

"No Dog" Device

A "no dog" device (**Figure 19**) is a bender accessory that contains a level or series of levels that helps establish the correct angles. Improper angles and alignment of conduit during bending can lead to "dog legs," or poor bends.

Figure 19 "No dog" device.

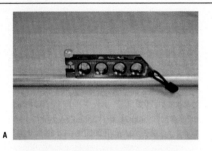

A

B

Courtesy of Alan W. Stanfield.

CONDUIT-BENDING MEASUREMENTS

Take-Up

Bender take-up is the length of conduit that is "taken up" in the bending of the conduit (**Table 10**). This amount should be subtracted from the original stub-up length when making bends.

Table 10   Conduit Bender Take-Up

Conduit Type and Trade Size	Bender Take-Up	
	in.	mm
½-in. EMT	5	127
¾-in. EMT or ½-in. rigid conduit/IMC	6	152
1-in. EMT or ¾-in. rigid conduit/IMC	8	203
1¼-in. EMT or 1-in. rigid conduit/IMC	11	279

Data from Integrated Publishing, Inc., *Conduit Bending* [http://www.tpub.com/ceb/68.htm]. Accessed August 11, 2010.

Gain

When bending a piece of conduit, you are actually cutting the corner off the angle. In other words, if you were to create a 90-degree bend in a piece of conduit and measure both horizontal and vertical portions, the actual length of the conduit would be less than the sum of the two portions. The gain is simply the difference between the sum of the two legs of the conduit and the actual length of the conduit (**Table 11**). The arc of the bend allows the conduit to be shorter than two straight pieces that run at 90 degrees.

CONDUIT-BENDING MEASUREMENTS

Table 11 90-Degree Hand Bender Gain for Rigid Conduit

Trade Size	Gain	Radius^a
½ in.	2⅝ in. (67 mm)	4 in. (102 mm)
¾ in.	3¼ in. (83 mm)	5 in. (127 mm)
1 in.	4 in. (102 mm)	6 in. (152 mm)
1¼ in.	5⅝ in. (143 mm)	8 in. (203 mm)

^aAccording to *NEC*® Chapter 9, Table 2.

Data from Bamford, B., *Bending Conduit/Tubing Using Hand Benders*—CEFGA
[http://www.cefga.org/documents/BendingRaceways.pdf]. Accessed August 19, 2010.

The actual gain will vary with the size of the conduit and the radius of the bend. Therefore, to more precisely determine the proper gain, each bender should be tested with a scrap piece of conduit. The amount of gain may also vary somewhat with each type of bender.

Pipe Expansion

As temperature increases, conduit can expand. This factor must be considered at all sites but particularly when installing long runs. PVC expands at a greater rate than metal conduits (**Table 12**). The *NEC*® provides information about conduit expansion in Table 352.44. As the distance of the conduit run increases and high temperatures are encountered, expansion fittings may be required to maintain the integrity of the conduit installation.

CONDUIT-BENDING MEASUREMENTS

Table 12   PVC Expansion

Temperature Change (°F)	Length Change (in./100 ft)	Temperature Change (°C)	Length Change (mm/m)
10	0.41	10	0.61
20	0.81	20	1.22
30	1.22	30	1.83
40	1.62	40	2.43
50	2.03	50	3.04
60	2.43	60	3.65
70	2.84	70	4.26
80	3.24	80	4.87
90	3.65	90	5.48
100	4.06	100	6.08

Note: Refer to *NEC*® Section 300.7 for the expansion coefficients of steel and aluminum.

Modified from *NEC*® Table 352.44.

NEC® Notes on Bends

The *NEC*® states in Sections 342.26, 344.26, 348.26, 353.26, 358.26, and 362.26 that the maximum number of bends between boxes or conduit bodies should not exceed the equivalent of four 90-degree bends, or 360 degrees total.

There is other information found in the *NEC*® that relates to the securing and supporting of the various types of conduit. Attention should also be paid to any reaming or threading requirements listed in the *NEC*® specific to each type of conduit. Most of this information can be found between Articles 342 and 362.

🔌 STUB-UP BENDS

This section describes how to make the most accurate stub-up bend. Stub-up bends are normally used in situations where a conduit running along the floor or ceiling requires a 90-degree bend to change the direction of the conduit. For example, a conduit running along a wall may need to turn inside a corner or exit up through a concrete slab floor. Stub-up bends are generally the easiest types of conduit bends to make.

Making a Stub-Up Bend Using a Hand Bender

Step 1: Determine the bender take-up based on the type and size of the conduit to be bent.

Referring to Table 10 (see page 30), if the conduit is ½-in. electrical metallic tubing (EMT), then bender take-up is 5 in. (127 mm). A piece of ¾-in. EMT requires a bender take-up of 6 in. (152 mm).

Step 2: Determine which size of stub-up bend is needed, and make the proper mark on the conduit.

If a 12-in. (305-mm) stub-up bend is needed on a piece of ½-in. EMT, you must subtract the 5-in. (127-mm) take-up from 12 in. (305 mm). The bender take-up subtracted from the original conduit length equals 7 in. (178 mm). Mark the conduit 7 in. (178 mm) from the end (**Figure 20**).

Figure 20   Step 2: Make the stub-up mark.

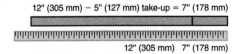

12" (305 mm) − 5" (127 mm) take-up = 7" (178 mm)

12" (305 mm)   7" (178 mm)

33

STUB-UP BENDS

If a 12-in. (305-mm) stub-up bend is needed on a piece of ¾-in. EMT, you must subtract the 6-in. (152-mm) take-up from 12-in. (305 mm). This would give you 6-in. (152 mm). The conduit would then be marked 6-in. (152 mm) from the end.

Step 3: Make the stub-up bend.

Once the stub-up mark has been made, insert the bender onto the conduit and place the conduit securely on the floor. The stub-up arrow should be placed on the stub-up mark (**Figure 21**).

Figure 21 Step 3A: Insert the bender onto the conduit.

Place one foot on the conduit and the other foot onto the bender foot pedal. Apply equal force with your hands and foot to ensure that the bender does not slip or rise (**Figure 22A**).

Continue bending the conduit until the 90-degree marking on the bender is parallel to the conduit. For example, if a 9-in. (229-mm) stub-up bend is needed for a piece of ½-in. EMT, you must subtract the 5-in. (127-mm) bender take-up from 9 in. (229 mm) (**Figure 22B**).

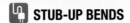
STUB-UP BENDS

Figure 22A Step 3B: Apply force to make the bend.

Figure 22B Step 3C: Continue the bend to form a 90-degree stub up.

Thus, you should mark the conduit 4 in. (102 mm) from the end. Ensure that the conduit remains in contact with the bender for the entire length of the bend (**Figure 23**).

Figure 23   There should be no gap between the bender and conduit.

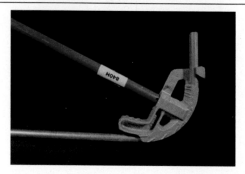

After you have completed the bend, position the conduit so that the end is facing up, and measure the distance from the floor to the end of the conduit. It should measure approximately 9 in. (229 mm) (**Figure 24**).

STUB-UP BENDS

Figure 24 Step 3D: Ensure that the bend is 9 in. (229 mm) from the floor.

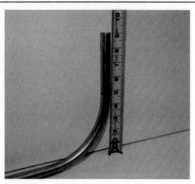

Courtesy of Alan W. Stanfield.

Making a Stub-Up Bend Using a Mechanical Bender

Step 1: Measure and mark the conduit.

Mark (A) for the desired height of the stub-up bend. For example, if a 12-in. (305-mm) stub-up bend is needed, place mark (A) 12-in. (305-mm) from the end of the pipe.

Next, use the chart located on the side of the bender to determine the distance of take-up for the bender. For the bender in this example, it is 5½ in. (140 mm) for a piece of ½-in. RMC.

Place mark (B) 5½ in. (140 mm) back toward the end of the conduit. In this example, mark (B) should be at 5½ in. (165 mm) from the end of the conduit (**Figure 25**).

STUB-UP BENDS

Figure 25   Step 1: Measure and mark the conduit.

Courtesy of Alan W. Stanfield.

Step 2: Make the stub-up bend.

Place the conduit into the bender with mark (B) located at the front edge of the hook (**Figure 26**).

Figure 26   Step 2A: Place the conduit into the bender.

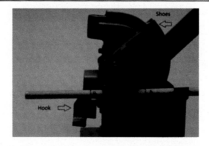

 **STUB-UP BENDS**

Pull the lever until the bend reaches 90 degrees (**Figure 27**).

Figure 27 Step 2B: Pull the lever until the bend reaches 90 degrees.

Courtesy of Alan W. Stanfield.

Stub-Up Bend Example

Example

An 18-in. (475-mm) stub-up bend is needed for a piece of ¾-in. EMT. Where should the conduit be marked in preparation for bending, and what should the stub-up height be?

Solution

A ¾-in. EMT has a take-up of 6 in. (152 mm). Mark the conduit 12 in. (305 mm) from the end and make a 90-degree bend. The height of the stub-up bend should be 18 in. (475 mm).

BACK-TO-BACK BENDS

Back-to-back bends are used in situations that require two 90-degree bends to be set at a predetermined distance apart. **Figure 28** demonstrates one type of back-to-back bend. In this figure, the two ends of the conduit point in the same direction. Another type of back-to-back bend could have one end of the conduit parallel to the floor while the other end is parallel to the wall (**Figure 29**). A back-to-back bend is made by marking the conduit twice and making two bends.

Figure 28   A standard back-to-back bend.

Courtesy of Alan W. Stanfield.

Figure 29   A back-to-back bend with one end parallel to the floor and the other end parallel to the wall.

Courtesy of Alan W. Stanfield.

🔌 BACK-TO-BACK BENDS

Making a Back-to-Back Bend

Step 1: Determine the height of the first 90-degree stub-up bend and the distance needed between the bends.

In some cases, the height of the stub-up bends may not matter too much, but the measurement of the distance between the bends will still be important. If this is the case, then first determine a height for the first stub-up bend. You will probably want at least a 12-in. (305-mm) stub-up bend to allow for room to make the bend. In this example, we will use a 30-in. (762-mm) distance between the two bends.

When making a back-to-back bend, it is important that you allow enough room for the handle and the shoe of the bender to clear the other bend. The shoe or head of the bender must have enough space to complete the radius of the bend without interfering with the other bend in the conduit. It is a good idea to avoid back-to-back bends that are less than 12 in. (305 mm).

Step 2: Make the first stub-up mark and then make the first stub-up bend.

In this example, we will use a 15-in. (381-mm) stub-up bend for the first bend. Make the first 15-in. (381-mm) stub-up bend (**Figure 30**). If using ½-in. EMT, you should subtract the 5-in. (127-mm) bender take-up from 15 in. (381 mm) and make the bend at 10 in. (254 mm) from the end of the conduit.

BACK-TO-BACK BENDS

Figure 30 Step 2: Make the first stub-up bend.

Step 3: Make the second stub-up mark, and then make the second stub-up bend.

To make the second bend, place a straight edge on the back of the first bend so that you can accurately measure from the back of the first bend; mark 30 in. (762 mm) on the other end (**Figure 31**). Now, using the star point, make the second bend back toward the first (**Figure 32A**). Complete the 90-degree bend to form the back-to-back bend (**Figure 32B**).

BACK-TO-BACK BENDS

Figure 31 Step 3A: Mark the conduit a given distance away from the first stub-up bend.

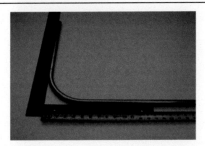

Courtesy of Alan W. Stanfield.

Figure 32A Step 3B: Make the second stub-up bend.

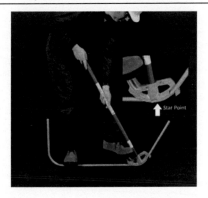

Figure 32B Step 3B: Complete the second stub-up bend.

If the distance between the two bends is smaller, such as 20 in. (508 mm), you probably want to make the second bend in the opposite direction. You can do this by subtracting the 5-in. (127-mm) take-up from 20 in. (508 mm) and marking the conduit at 15 in. (381 mm). The bend would then be made by using the stub-up arrow and bending away from the first bend (**Figure 33A**). Complete the 90-degree bend to form the back-to-back bend (**Figure 33B**).

BACK-TO-BACK BENDS

Figure 33A Step 3C: Make the second stub-up bend in the opposite direction.

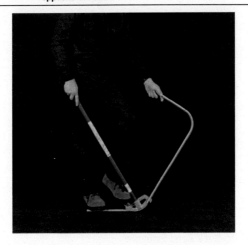

BACK-TO-BACK BENDS

Back-to-Back Bend Example

Example

A 36-in. (914-mm) back-to-back bend is needed for a piece of ½-in.
EMT. Use a 12-in. (305-mm) stub-up bend for the first bend.

Solution

Make a 12-in. (305-mm), 90-degree stub-up bend on one end of the
conduit. Measure from the back of the first bend 36 in. (914 mm) and
mark the conduit. Make the second bend using the star point, also
known as the back-of-bend mark, on the bender. The result should be a
36-in. (914-mm) back-to-back bend.

OFFSET BENDS

Offset bends are very common in the electrical field. They are used to enter boxes or cabinets at a direct angle to ensure proper alignment into fittings. Proper offset bending leads to good mechanical connections as well as neat workmanship. In addition, offset bends are used to direct conduit over obstacles such as pipes, beams, or other objects that may obstruct it from running in a straight line (**Figure 34**).

Figure 34 An offset bend.

7" (178 mm)

An offset bend is created by making two marks on the conduit and forming two bends. You will need to make calculations using a formula and allow for shrink of the conduit.

Using Trigonometry to Make Offset Bends

Because an offset requires two bends to complete, you must determine the distance between the two bends. This is accomplished by first determining the most appropriate angles to use for the offset, such as 30 degrees, 45 degrees, or 60 degrees. Next, locate the corresponding sine (SIN) in **Table 13**.

 **OFFSET BENDS**

Table 13   Most Common Sine Angles Used to Make Offset Bends

Angle	Sine
5°	0.0872
10°	0.1736
22½°	0.3827
30°	0.5000
45°	0.7071
60°	0.8660

Using the sine of the angle, determine the angle's cosecant (CSC). The CSC is the multiplier used along with the offset height to determine the distance between the bends of the offset.

For example, if a 6-in. (152-mm) offset is needed and 30-degree angles are to be used, locate the sine of 30 degrees on the trigonometry table: 0.5000. Next, use the following formula to determine the CSC or multiplier:

$$CSC(30°) = 1/SIN(30°) = 1/0.5000 = 2$$

In this example, the CSC (multiplier) is 2.

Calculating the Distance Between the Bends of an Offset

Once you have determined the correct multiplier, multiply the CSC—2, in our example—by the offset height of 6 in. (152 mm) to determine the distance between the bends (**Figure 35**).

Distance between the bends = 2 × 6 in. (152 mm) = 12 in. (305 mm)

OFFSET BENDS

Figure 35 The distance between the bends using 30-degree angles is 2× the offset height.

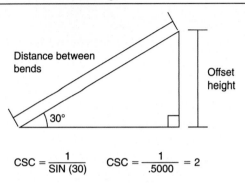

$$CSC = \frac{1}{SIN\ (30)} \qquad CSC = \frac{1}{.5000} = 2$$

The distance between the bends using 30-degree angles is 2× the offset height.

If 45-degree angles are used, the formula would be as follows:

$$CSC(45°) = 1/SIN(45°) = 1/0.7071 = 1.414$$

Distance between the bends $= 1.414 \times 6$ in. (152 mm)

$$= 8.48\ \text{in. (215 mm)}$$

Tables 14 to **18** serve as a quick reference for determining the distance between the bends as well as the shrink for a few sample offsets. The data in these tables have been calculated using the formulas found in **Table 19**. Notice how the smaller height offsets use the 22½-degree and 30-degree angles. Generally, for objects 2 in. (51 mm) to 4 in. (102 mm) in size, it is best to use 22½-degree or 30-degree angles. For objects that are 5 in. (127 mm) or larger, it is best to use 45-degree or 60-degree angles, as shown in Tables 17 and 18. Because each offset bend scenario can be different, however, you should use the best angle that most appropriately and safely fits the job.

 **OFFSET BENDS**

Table 14 Distance Between Bends and Shrink for
10-Degree × 10-Degree Offset Bends

Offset Height		Distance Between Bends		Shrink (per in.)	
in.	mm	in.	mm	in.	mm
2	51	11½	292	⅛	3
3	76	17¼	438	3/16	5
4	102	23	584	¼	6
5	127	28¾	730	5/16	8
6	152	34½	876	⅜	10
7	178	40¼	1022	7/16	11
8	203	46	1168	½	13

Note: Calculations under Distance Between Bends are rounded to the nearest ¼ in.

Table 15 Distance Between Bends and Shrink for
22½-Degree × 22½-Degree Offset Bends

Offset Height		Distance Between Bends		Shrink (per in.)	
in.	mm	in.	mm	in.	mm
2	51	5¼	133	⅜	10
3	76	7¾	197	9/16	14
4	102	10½	267	¾	19
5	127	13	330	15/16	24
6	152	15½	394	1⅛	29
7	178	18¼	464	15/16	33
8	203	20¾	527	1½	38

Note: Calculations under Distance Between Bends are rounded to the nearest ¼ in.

 **OFFSET BENDS**

Table 16 Distance Between Bends and Shrink for
30-Degree × 30-Degree Offset Bends

Offset Height		Distance Between Bends		Shrink (per in.)	
in.	mm	in.	mm	in.	mm
2	51	4	102	½	13
3	76	6	152	¾	20
4	102	8	203	1	25
5	127	10	254	1¼	32
6	152	12	305	1½	38
7	178	14	356	1¾	44
8	203	16	406	2	51

Note: Calculations under Distance Between Bends are rounded to the nearest ¼ in.

Table 17 Distance Between Bends and Shrink for
45-Degree × 45-Degree Offset Bends

Offset Height		Distance Between Bends		Shrink (per in.)	
in.	mm	in.	mm	in.	mm
2	51	3	76	¾	19
3	76	4¼	108	1⅛	29
4	102	5¾	146	1½	38
5	127	7	178	1⅞	48
6	152	8½	216	2¼	57
7	178	9¾	248	2⅝	67
8	203	11¼	286	3	76

Note: Calculations under Distance Between Bends are rounded to the nearest ¼ in.

 **OFFSET BENDS**

Table 18 Distance Between Bends and Shrink for
60-Degree × 60-Degree Offset Bends

Offset Height		Distance Between Bends		Shrink (per in.)	
in.	mm	in.	mm	in.	mm
2	51	2¼	57	1	25
3	76	3½	89	1½	38
4	102	4½	114	2	51
5	127	5¾	146	2½	64
6	152	7	178	3	76
7	178	8	203	3½	89
8	203	9¼	235	4	102

Note: Calculations under Distance Between Bends are rounded to the nearest ¼ in.

Table 19 Offset Formula Chart

		Shrink (per in.)	
Angles Used	Multiplier of Offset Depth	in.	mm
10° × 10°	5.76	¹⁄₁₆	2
22½° × 22½°	2.61	³⁄₁₆	5
30° × 30°	2.00	¼	6
45° × 45°	1.41	⅜	10
60° × 60°	1.15	½	13

🔌 OFFSET BENDS

Making an Offset Bend Using a Hand Bender

Step 1: Determine the height of the object.

The first task is to determine the size of the object, as it will determine the distance between the marks on the conduit. If the object is 7 in. (178 mm) high, for example, you would use ½-in. EMT.

Step 2: Determine the proper angle and calculate the distance.

A variety of angles can be used when creating offset bends. It is recommended that you use an angle that is neither too large nor too small. A 45-degree angle is often too large because it can make pulling the conductors difficult. A very small angle, such as one that is 10 degrees, produces an offset bend that requires a lot of room and can be unsightly. The best angle to use is a 30-degree angle. This angle is easier for wire pulling but is also a compact angle that forms a nice fit over the object. In some situations, however, a 10-degree, 22½-degree, 45-degree, or 60-degree angle may be a better alternative.

After determining the most appropriate angle, use the offset formula chart in Table 19 (see page 52). Using a 7-in. (178-mm) offset for this example and a 30-degree angle, you would multiply the offset height by the multiplier to achieve the distance between the two bends.

7 in. (178 mm) × 2 = 14 in. (356 mm) between the two marks

If you had an object with a height of 10 in. (254 mm) and were using an angle of 22½ degrees, you would calculate the distance as follows:

10 in. (254 mm) × 2.61 = 26.1 in. (663 mm)

which converts to approximately 26⅛ in. (664 mm) between the two marks. Typically, it is appropriate to round the decimal numbers to the nearest ⅛ in. or ¼ in.

OFFSET BENDS

Step 3: Make the marks and determine the shrink.

When an offset is bent on a conduit, the conduit will shrink in length due to the angles used. In this example, use a 7-in. (178-mm) offset with 30-degree angles. According to Table 19, this choice would produce a shrink of ¼ in. (6 mm) per inch of offset height.

7 in. × ¼ in. = 1¾-in. (44-mm) shrink

In the example, there is a 7-in. (178-mm) offset bend using 30-degree angles, and there is a shrink of 1¾ in. (44 mm).

Now that the calculations are complete, the conduit is ready to be marked. Suppose that the 7-in. (178-mm) object is 40 in. (1016 mm) from the end of the last coupling on the conduit run. You must account for the shrink, so the first mark (A) on the conduit would be 41¾ in. (1060 mm) from the end of the conduit.

The second mark (B) would be 14 in. (356 mm) from mark (A), which was calculated in Step 2. Notice that mark (B) is made toward the end where the first measurement was made, which is to the left of the first mark (A) (**Figure 36**).

Figure 36   Step 3: Mark the conduit 14 in. (356 mm) from mark (A).

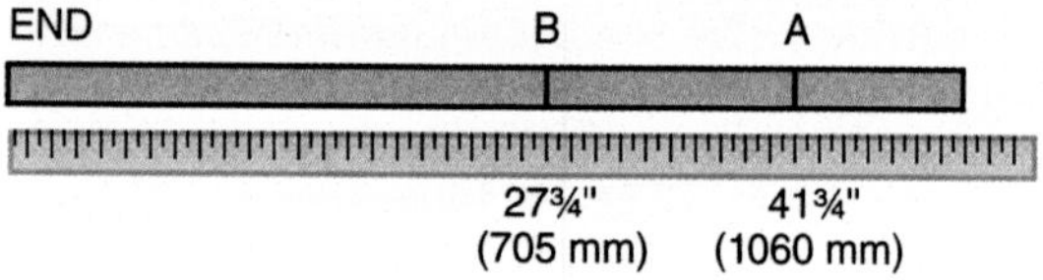

END		B	A	

27¾"
(705 mm)

41¾"
(1060 mm)

🔧 OFFSET BENDS

Another helpful tip is to always mark the end of the conduit from which all of the measurements are made. You can also use a coupling to represent the reference end. This practice will keep you from getting confused about which direction to bend the conduit in Step 4.

40 in. (1016 mm) + 1¾-in. (44-mm) shrink = 41¾ in. (1060 mm)

Step 4: Make the first bend.

Now that the conduit is properly marked, you can make the first bend.

Place the stub-up arrow on mark (A) and bend the conduit back toward the end mark or coupling end to 30 degrees (**Figure 37**). One of the problems encountered when bending offsets is overbending or underbending the conduit. It can be somewhat difficult to see if the angle is totally accurate while bending. However, 30-degree angles are a little easier to determine because the bender handle is 90 degrees to the floor when the 30-degree point is reached (**Figure 38**).

Figure 37 Step 4: Bend the conduit back toward the end mark to 30 degrees.

OFFSET BENDS

Figure 38 Bender is 90 degrees to the floor.

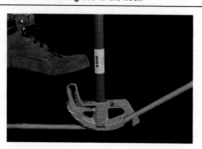

Step 5: Make the second bend.

To complete the second bend, leave the conduit in the bender.

Pick up the conduit and bender. Turn both upside down, so that the bender handle is placed on the floor.

Slide the bender back to the mark (B) using the stub-up arrow.

Rotate the conduit in the bender 180 degrees (**Figure 39**). Make sure that the conduit is lined up with the bender and that the second bend is made with the conduit rotated 180 degrees. If the conduit is off by more than a couple of degrees, it will produce a crooked offset.

When the conduit is properly aligned, place the end under your armpit and apply pressure to the conduit.

Keep one foot securely placed behind the bender handle to stabilize the bender and keep it from slipping.

Bend the conduit to 30 degrees (**Figure 40**). The conduit should fit over the 7-in. (178-mm) object (**Figure 41**).

OFFSET BENDS

Figure 39 Step 5A: Rotate the conduit in the bender 180 degrees.

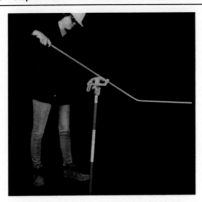

Figure 40 Step 5B: Bend the conduit to 30 degrees.

Figure 41 Step 5C: Make sure the conduit fits over the object.

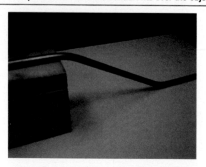

Making an Offset Bend Using a Mechanical Bender

Step 1: Mark the conduit.

Begin by placing mark (A) at a predetermined distance from the end of the conduit.

Using the table on the bender, determine the distance between the two bends. For example, a 6-in. (152-mm) offset using 30-degree angles would require 12 in. (305 mm) between the bends. For this example, place mark (B) at a point 12 in. (305 mm) from mark (A) using 30-degree angles (**Figure 42**). Using 15-degree bends and 40-degree bends would require different measurements between the bends. Refer to the chart on the bender for more details. In general, 30-degree angles are easier to use because the multiplier for the distance between the bends is 2. Therefore, a 4-in. (102-mm) offset using 30-degree angles would result in 8 in. (203 mm) between the bends.

⚡ OFFSET BENDS

Figure 42   Step 1: Mark the conduit.

Step 2: Make the first bend.

Place the conduit into the bender with mark (A) on the front of the hook.

Bend the conduit to the desired angle. For this example, 30-degree angles are used (**Figure 43**).

Figure 43   Step 2: Make the first bend.

OFFSET BENDS

Step 3: Make the second bend.

Slide the conduit down to mark (B) and rotate it 180 degrees.

With mark (B) aligned to the front edge of the hook, make the second
bend to the desired angle. For this example, the angles used are
30 degrees (**Figure 44**).

Figure 44 Step 3: Make the second bend.

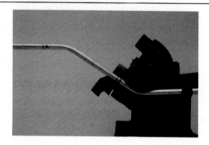

Offset Bends for Entering Boxes

Where a conduit enters a box or cabinet, a small offset of 2 in. (51 mm)
or less may be required (**Figure 45**). In this situation, it is best to use
a 10-degree or 22½-degree angle for the bends. Often, these small
offsets are made without any measuring or calculations. For example,
the quickest way to make an offset of a 1 in. (25 mm) or less is to make
the first bend approximately 2 in. (51 mm) from the end of the conduit
that enters the box. Make this first bend on the floor using a 10-degree
angle. Next, pick up the conduit and the bender, and turn both upside
down. Slide the conduit forward, away from the end, entering the box
approximately 1 to 2 in. (25 to 51 mm). Finally, rotate the conduit 180
degrees and make the second bend using a 10-degree angle.

 OFFSET BENDS

Figure 45 A small offset in a pipe attached to a box.

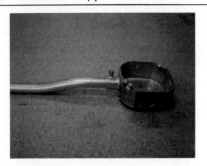

Rolling Offsets

To determine how much offset is needed to make a rolling offset:

1. Measure the vertical distance required. Use the worktable (any square will do) and measure this length (distance calculated) from a corner, then mark the worktable or square.

2. Measure the horizontal distance required. Measure 90 degrees from the vertical line measurement (starting in same corner) and mark the worktable or square.

3. The diagonal distance between these marks will be the amount of offset required.

Note: The shrink is calculated as the hypotenuse minus the side adjacent to the angle.

Parallel Offsets

Another common type of bend is a series of parallel offset bends. This type of bend is often necessary when several different runs of conduit are placed in parallel with others and an offset is required to clear an object or necessary for the conduits to enter an area that may be offset (**Figure 46**).

Figure 46   Three parallel offset runs.

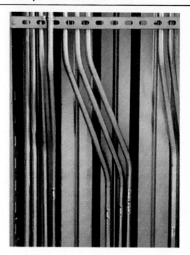

The common problem with parallel offsets is that if the conduits are equally spaced, it will result in crowding along the angled section of the bend. Therefore, it is necessary to adjust or shift the bend marks for each adjacent conduit. The following steps demonstrate how to perform 30-degree parallel offsets:

1. Make the first 30-degree offset bend as normal.
2. Using the trigonometric functions table (pp. 126–127), take the 30-degree angle and divide it by 2, which results in 15 degrees.

⚡ OFFSET BENDS

3. Locate the 15-degree tangent on the table, which is 0.2679; for simplicity, reduce the number to 0.27 to make the calculation easier. This number becomes the constant by which to multiply the distance.

4. Multiply the constant 0.27 by the desired distance between the bends to achieve the adjustment measurement. For example, if the desired spacing between the offsets is 2 in., then multiply 0.27 by 2 (0.27×2 in.) to equal an adjustment of 0.54 in.

5. Adjust or shift the conduit marks down on the second offset bend by 0.54 in. (round to ½ in.). Continue this process of shifting the marks down by ½ in. for each consecutive offset bend (**Figure 47**).

Thirty-degree offsets are the easiest to use when calculating parallel offsets because the constant is ½ in. Other offset angles can be used to achieve the same result but will require referring to the trigonometric functions table (pp. 126–127) to obtain the tangent. The process is the same for an angle chosen.

Figure 47 Marks on the conduit are moved down by ½ in.

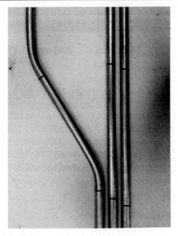

Courtesy of Alan W. Stanfield.

OFFSET BENDS

Offset Bend Examples

Example 1

Create an offset bend for an object that is 8 in. (203 mm) high using a 45-degree angle. The object is 24 in. (610 mm) from the last coupling. Use ½-in. EMT.

Solution

Multiply 8 in. (203 mm) by 1.41 (= 11.28), which, converted to a fraction, is approximately 11¼ in. (286 mm); thus, the distance between the bends is 11¼ in. (286 mm). The shrink is ⅜ in. (10 mm) per inch: 8 in. × ⅜ in. = 3 in. (76 mm). Mark (A) should be 27 in. (686 mm) from the coupling. Mark (B) should be 11¼ in. (286 mm) from mark (A) and should be measured back toward the coupling (**Figure 48**).

Figure 48 Marks (A) and (B) for offset bend example 1.

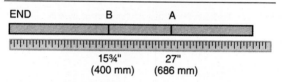

Example 2

Create a 2-in. (51-mm) offset bend using 22½-degree angles. The object is 28 in. (711 mm) from the last coupling. Use ½-in. EMT in this practice bend.

Solution

Refer to Table 15 (see page 50). Mark (A) should be located at 28¾ in. (730 mm) from the coupling. Mark (B) should be at 23½ in. (597 mm) from the coupling.

🔌 THREE-BEND SADDLES

Three-bend saddles are useful when you need to clear a water pipe, vent pipe, or any other round object that may be in the path of the conduit run. They are normally used for objects that are less than 7 in. (178 mm) in size. Although three-bend saddles can be made with 30-degree and 60-degree bends, the best angles to use are typically 45-degree and 22½-degree angles.

Making a Three-Bend Saddle

Step 1: Determine the distance to the object and the height.

First, determine how far the center of the object is from the last coupling on the conduit. Next, determine the height of the object. These measurements will provide the necessary information for making a proper bend. In this example, the object is 36 in. (914 mm) from the coupling and is 4 in. (102 mm) high. Use ½-in. EMT (**Figure 49**).

Figure 49　Step 1: Determine the distance to the object and the object's height.

36" (914 mm)

4"
(102 mm)
high

Step 2: Determine the angles and perform the calculations.

Three-bend saddles can be made with 30-degree and 60-degree bends or 22½-degree and 45-degree bends. The 45-degree–22½-degree combination provides a smoother appearance and is easier to pull wire through. This example uses the 45-degree–22½-degree combination.

THREE-BEND SADDLES

Once the angle has been determined, the distance between the bends and the shrink amount need to be calculated. To determine the shrink for 22½-degree and 45-degree bends, note that the shrink amount is 3/16 inch (5 mm) for every 1 inch (25 mm) of the object's height. To determine the distance between the bends, use the following formula:

Distance between bends = diameter of the object × 2½

For example, if the object is 4 in. (102 mm) high, the distance between the bends would be 4 in. (102 mm) × 2½ = 10 in. (254 mm), and the shrink would be 4 in. × 3/16 in. = ¾ in. (19 mm). **Table 20** provides a quick reference for some common obstruction sizes.

Table 20 Three-Bend Saddles Using One 45-Degree and Two 22½-Degree Bends

Height of Object		Shrink (per in.)		Distance from Center Bend	
in.	mm	in.	mm	in.	mm
1	25	3/16	5	2½	64
2	51	3/8	10	5	127
3	76	9/16	14	7½	191
4	102	¾	19	10	254
5	127	15/16	24	12½	318

Note: Calculations under Distance from Center Bends are rounded to the nearest ¼ in.

Step 3: Mark the conduit.

The next step is to mark the conduit. Three-bend saddles should have three marks made on the conduit.

The first mark (A) is the distance from the coupling to the center of the object, considering shrink. In this example, mark (A) would be placed at 36 in. (914 mm) + ¾ in. (19 mm) shrink = 36¾ in. (933 mm).

THREE-BEND SADDLES

Marks (B) and (C) would be made 10 in. (254 mm) from mark (A). Place mark (B) 10 in. (254 mm) to the left of (A) and mark (C) 10 in. (254 mm) to the right of (A) (**Figure 50**).

Figure 50   Step 3: Mark the conduit.

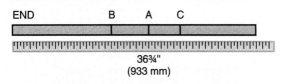

Step 4: Make the first bend.

The first bend is made at mark (A) using the teardrop or notch on the bender. These teardrop symbols or notches represent the back of the 45-degree bend.

Place the conduit into the bender with mark (A) aligned with the teardrop or notch.

Bend the conduit to 45 degrees on the floor (**Figure 51**).

Figure 51   Step 4: Make the first bend.

67

THREE-BEND SADDLES

Step 5: Make the second and third bends.

The second and third bends are 22½-degree bends. They should be made using the stub-up arrow.

After you have completed the first bend, pick up the bender and conduit, and turn them upside down. The bender handle should now be on the floor (**Figure 52**).

Figure 52   Step 5A: Make the second bend.

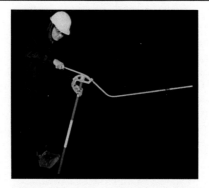

Slide the conduit to mark (B) and, using the stub-up arrow, rotate the conduit 180 degrees.

Place the end of the conduit under your armpit and bend it down to 22½ degrees. Make sure that the conduit remains straight when you are making the second and third bends. The conduit should fit within ¼ in. (6 mm) of the object.

Make the third bend by removing the conduit from the bender. Place mark (C) on the stub-up arrow and make a 22½-degree bend (**Figure 53**). The conduit should fit within ¼ in. (6 mm) of the object (**Figure 54**).

THREE-BEND SADDLES

Figure 53   Step 5B: Make the third bend.

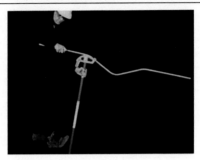

Figure 54   Step 5C: A completed three-bend saddle.

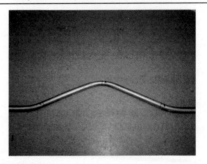

Courtesy of Alan W. Stanfield.

THREE-BEND SADDLES

Making a Three-Bend Saddle Using 30-Degree and 60-Degree Angles

If 30-degree and 60-degree bends are used, the shrink is ¼ in. (6 mm) per inch of height; the distance between the bends is two times the height of the object. For example, if the object is 3 in. (76 mm) high, the shrink would be ¾ in. (19 mm), and the distance between the bends would be 6 in. (152 mm). The 30-degree–60-degree combination is easier to calculate but does not always offer the most suitable bend; therefore, the 45-degree–22½-degree combination is used most often. **Table 21** provides a quick calculation reference when using 30-degree and 60-degree angles.

Table 21   Three-Bend Saddles Using One 60-Degree and Two 30-Degree Bends

Height of Object		Shrink (per in.)		Distance from Center Bend	
in.	mm	in.	mm	in.	mm
1	25	¼	6	2	51
2	51	½	13	4	102
3	76	¾	19	6	152
4	102	1	25	8	203
5	127	1¼	32	10	254

Note: Calculations under Distance from Center Bends are rounded to the nearest ¼ in.

Three-Bend Saddle Example

Example

Bend a three-bend saddle for an object that is 5 in. (127 mm) high and is located 48 in. (1219 mm) from the last coupling using 45-degree and 22½-degree angles. Use ½-in. EMT.

Solution

The shrink of the conduit would be 5 in. × ³⁄₁₆ in. = `1 (24 mm). The distance from mark (B) to mark (A) and from mark (C) to mark (A) would be 5 in. (127 mm) × 2½ = 12½ in. (318 mm). The conduit would be marked with mark (A) at 48¹⁵⁄₁₆ (1243 mm), mark (B) at 36⁷⁄₁₆ in. (926 mm), and mark (C) at 61⁷⁄₁₆ in. (1561 mm).

70

FOUR-BEND SADDLES

Four-bend saddles are used when a square or rectangular object is encountered on a conduit run. A four-bend saddle is really just two offset bends that are combined to form one smooth transition over an object (**Figure 55**). The same principles used in making an offset bend are needed, and the offset guide table (see Table 19, page 52) can be used to make this type of bend.

Figure 55 A completed four-bend saddle used to transition over an object.

Making a Four-Bend Saddle

Step 1: Determine the height and width of the object.

Measure the height and width of the object or obstruction. Also measure the distance from the last coupling to the front edge of the object. Use ½-in. EMT (**Figure 56**). In this example, the object is 10 in. (254 mm) wide, 4 in. (102 mm) high, and 30 in. (762 mm) from the last coupling.

FOUR-BEND SADDLES

Figure 56 Step 1: Determine the height and width of the object.

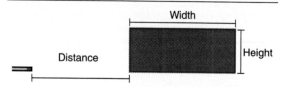

Step 2: Determine the angles and perform the calculations.

Four-bend saddles are made in the same way as offset bends, except that two offsets are needed to complete the bend. Although you can use a variety of angles, be aware that 30-degree bends tend to work best. In this example, 30-degree bends will be used. Table 19 (see page 52) will be used to determine the measurements for this bend.

Because 30-degree angles are being used in this example, the shrink would be 1 in. (4 in. × ¼ in. = 1 in. [25 mm]). The distance between the bends would be 8 in. (4 in. × 2 = 8 in. [203 mm]).

Step 3: Mark the conduit.

Now that the calculations have been performed, you can mark the conduit. The conduit will have four marks. These marks start with (A), which is the closest to the last coupling.

Because the object is 30 in. (762 mm) from the last coupling, mark (A) should be at 31 in. (30 in. + 1 in. shrink = 31 in. [787 mm]).

Mark (B) should be 8 in. (203 mm) from mark (A).

The third mark (C) is determined by the width of the object, which in this case is 10 in. Mark (C) should be 10 in. (254 mm) from mark (B).

🔌 FOUR-BEND SADDLES

Mark (D) should be the same as the distance between mark (A) and mark (B), which is 8 in. (203 mm).

In summary, mark (A) should be made at 31 in. (787 mm), mark (B) at 39 in. (991 mm), mark (C) at 49 in. (1245 mm), and mark (D) at 57 in. (1448 mm) (**Figure 57**).

Figure 57   Step 3: Mark the conduit.

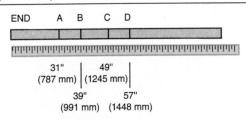

Step 4: Make the first bend.

Recall that four-bend saddles are made by bending two offsets opposite of each other. Just as with an offset bend, the first bend is actually made at mark (B).

Place stub-up arrow on mark (B) and make a 30-degree bend. You can make this first bend on the floor (**Figure 58**).

Figure 58   Step 4: Make the first bend.

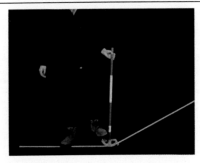

Step 5: Make the second bend.

The second bend is made by picking up the conduit and bender and turning them upside down, with the bender handle placed on the floor.

Slide the conduit to mark (A) and rotate it 180 degrees. Use the stub-up arrow for this bend, just as you would for a regular offset bend.

Place the conduit under your arm and bend it to 30 degrees (**Figure 59**).

Figure 59 Step 5: Make the second bend.

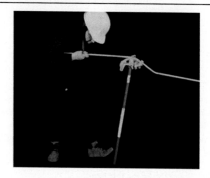

Step 6: Make the third bend.

Remove the conduit from the bender. Turn the conduit around so that the first two bends made are now pointing away from you.

Put the conduit back into the bender and slide it to mark (C) using the stub-up arrow. Place the conduit under your arm and bend it to 30 degrees (**Figure 60**).

Figure 60 Step 6: Make the third bend.

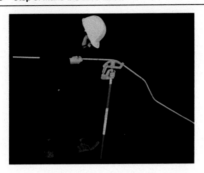

Step 7: Make the fourth bend.

Leave the conduit in the bender. Slide it to mark (D) using the stub-up arrow.

Rotate the conduit 180 degrees and align the conduit with the other bends to ensure a straight bend.

Place the conduit under your arm and bend it to 30 degrees (**Figure 61**). The conduit should fit over the 4-in. (102-mm) object (**Figure 62**).

FOUR-BEND SADDLES

Figure 61 Step 7A: Make the fourth bend.

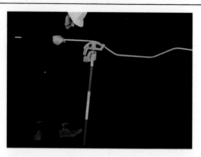

Figure 62 Step 7B: A completed four-bend saddle.

Four-Bend Saddle Example

Example

Create a four-bend saddle that will fit over an object that is 8 in. (203 mm) high and 12 in. (305 mm) wide. The distance to the last coupling is 24 in. (610 mm). Use 30-degree angles for this saddle. Use ½-in. EMT in this practice bend.

Solution

The shrink is 8 in. × ¼ in. = 2 in. (51 mm). The distance between mark (A) and mark (B) is 8 in. (203 mm) × 2 = 16 in. (406 mm). This is also the distance between mark (C) and mark (D). The distance between mark (B) and mark (C) should be 12 in. (305 mm). The conduit should be marked with mark (A) at 26 in. (660 mm), mark (B) at 42 in. (1067 mm), mark (C) at 54 in. (1371 mm), and mark (D) at 70 in. (1778 mm).

COMPOUND 90-DEGREE BENDS

Compound 90-degree bends are common in situations where a conduit enters a corner where an object such as a pipe or other obstruction does not permit the conduit to follow the contour of the corner (**Figure 63**). The compound 90-degree bend uses two bends to clear the object as opposed to a traditional 90-degree bend with only one angle.

Figure 63 Conduit bent to clear corner object.

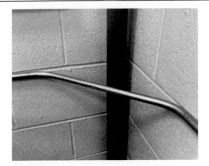

Courtesy of Alan W. Stanfield.

There is an easy way to perform compound 90-degree bends by using a measuring tape or rule to determine the distance between the bends. The most common bends used to form the compound 90-degree bend are 45-degree bends. The following steps demonstrate how to perform a compound 90-degree bend around a pipe.

COMPOUND 90-DEGREE BENDS

1. First, use a measuring tape or rule to determine the distance between the two bends. Measure from a straight line in front of the object and to the point where the conduit angles will contact the two walls (**Figure 64**).

Figure 64 Measure the distance between the two bends.

Courtesy of Alan W. Stanfield.

2. Determine the distance between the two bends and mark it on the conduit.

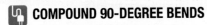

COMPOUND 90-DEGREE BENDS

3. Using the center bend mark (rim notch) on the conduit bender, bend the first 45-degree angle (**Figure 65**).

Figure 65 Bend the first 45-degree angle.

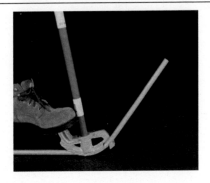

COMPOUND 90-DEGREE BENDS

4. Finally, place the center bend mark (rim notch) on the second bend with the conduit bender upside down and bend the second 45-degree angle (**Figure 66**).

Figure 66   Bend the second 45-degree angle.

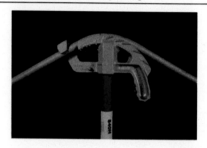

🔌 SEGMENTED BENDS

Making a Segmented Bend

Segmented bends combine a series of smaller bends, made using a multiple-shot hydraulic bender or a manual hickey bender. Segment benders require that the operator determine the degree and the number of smaller bends needed to complete the total desired bend on the conduit. For example, if a 90-degree angle is desired, then 18 bends at 5 degrees each can be made to form the desired 90-degree bend. Another example would be making nine bends at 10 degrees to form the desired total of 90 degrees. However, it should be noted that applying more bends to the conduit will produce a smoother appearance to the bend. Therefore, most segmented conduit bends include a greater number of bends at smaller angles for a better appearance. Fewer bends can make corrections to the conduit more difficult as well.

Steps to make a segmented bend:

1. Determine the radius for the desired bend. The easiest way to determine the radius of an object that the conduit must bend around is to find the diameter and divide by 2. For example, if the diameter of a tank that the conduit must bend around is 48 in., then the radius of the bend is 24 in.

2. Determine the developed length of the conduit, which is the total length of conduit required to make the bend. The formula for determining the developed length is:

    Developed length (DL) = Bend angle × radius × 0.0175

    For example: DL = 90° × 24 in. × 0.0175 or 37.8 in.

3. Determine the number of segment bends. If a 90-degree bend is desired, then determine the length of bends and appropriate angles to achieve 90 degrees. For example, 18 bends at 5 degrees will produce a 90-degree bend.

4. The final step is to determine the amount of space between the bends. This is done by dividing the developed length by the number of bends. For example:

    37.8 in. ÷ 18 = 2.1 in.

SEGMENTED BENDS

5. Mark the conduit. First, mark the conduit with two marks 37.8 in. apart. These are the two marks representing the developed length. Second, mark the center of the developed length. If the number of bends is an odd number, then place a mark to the right and left of the center mark at the determined spacing distance. If the number of bends is even, then the center mark represents the middle of a spacing measurement (**Figure 67**).

6. Make the bends. Remember that the two outside marks are not bending marks but only serve as the area of the developed length.

Figure 67 Even-numbered marks have a centerline between the middle segment.

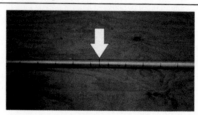

Segmented Bend Example

Example

Using a multiple-shot hydraulic bender, you wish to make a 45-degree bend in 9-in. (229 mm) shots. How many bends will you need to make, and what would be the required angle for each bend?

Solution

You will need to make nine bends at 5 degrees to create a 45-degree segmented bend.

Making a Concentric Bend

Concentric bends are bends that are made with multiple pieces of conduit run parallel to each other, to form an even and stacked appearance. These bends share a common center point, but the radius of each conduit becomes larger as each additional conduit is stacked onto the others (**Figure 68**).

CONCENTRIC BENDS

Figure 68   Concentric bends.

$$L + H + DL - 2R = NIPPLE$$

$$1.57 \times R = DL$$

$$H - R = \text{"B"}$$

$$L - R = \text{"C"}$$

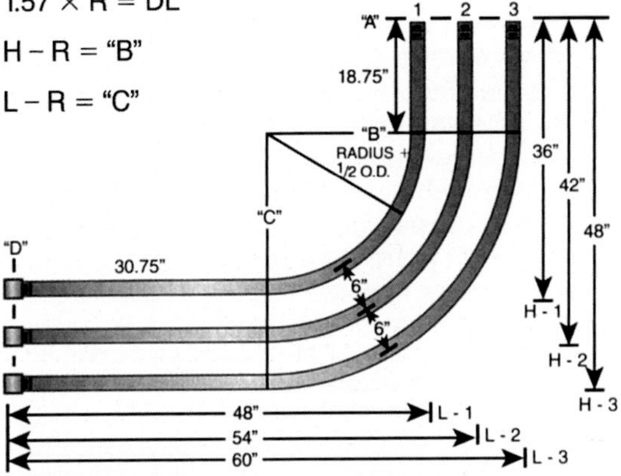

TO LOCATE POINT "B"

H#1 − RADIUS #1 = "B"
36" − 17.25"   = "B"
        18.75"   = "B"

H#2 − RADIUS #2 = "B"
36" − 23.25"   = "B"
        18.75"   = "B"

H#3 − RADIUS #3 = "B"
48" − 29.25"   = "B"
        18.75"   = "B"

TO LOCATE POINT "C"

L#1 − RADIUS #1 = "C"
48" − 17.25"   = "C"
        30.75"   = "C"

L#2 − RADIUS #2 = "C"
54" − 23.25"   = "C"
        30.75"   = "C"

L#3 − RADIUS #3 = "C"
60" − 29.25"   = "C"
        30.75"   = "C"

Points "B" and "C" are the same distance from the end on all three conduits.

CONCENTRIC BENDS

The center of each bend will remain the same, but the radius of each will change as each additional bend is added from the first. The radius of each bend will become larger for each conduit as it extends out from the first conduit. As this description suggests, making concentric bends can be complex and requires accurate calculations.

Steps to make concentric bends:

1. Determine the radius of the first desired conduit bend. For example, the first bend requires a radius of 30 in. (762 mm)

2. Determine the outside diameter (OD) of the conduit used. For example, if 3-in. (76 mm) conduit is used, the OD is 3.5 in. (89 mm)

3. Determine the distance or spacing between the conduits. For example, 2-in. (51 mm) spacing can be used.

4. After the OD and spacing are determined, then the centerline for each conduit can be found by adding the OD and spacing together for each conduit. For example, if the radius for the first conduit is 30 in., the second conduit would have a centerline of 30 in. + 3.5 in. + 2 in. = 35.5 in. The third conduit would have a centerline of 35.5 in. + 3.5 in. + 2 in. = 41 in. These centerline calculations will be used later to determine the distance between the bends for each conduit.

5. Bend the first conduit by determining the radius and the distance between the bends. For example, use a 3-in. conduit with a 30-in. radius and (15) 6-degree bends. Use the 30-in. radius and multiply by the tangent of 6 degrees, which is 0.105, to find the number of bends required.

$$30 \text{ in.} \times 0.105 = 3.15 \text{ in. between bends}$$

6. All preceding bends would be made by following the same steps, but the centerline measurement would change as described in Step 4. For example, the second concentric bend would be 35.5 in. × 0.105 = 3.725 in. Therefore, the distance between the bends for the second conduit would be 3.725 in., which would result in a larger radius bend.

CONCENTRIC BENDS

Concentric Bend Example

Example

Creating concentric bends can be difficult due to the spacing between the conduits, the need to use different sizes of conduit, and the requirement for performing calculations when several conduits are stacked together. A simple example would include stacking only two conduits together to form a concentric bend. In this example, the first conduit is bent using 18 shots at 5 degrees each to form a 90-degree angle. The second conduit would require how many shots, at what degree, and with how much spacing between the shots?

Solution

The second conduit would also require 18 shots at 5 degrees each. However, these shots would have to be separated at a greater distance than the first conduit. This separation between the shots may be ¼ in. (6 mm) or greater depending on how much space is desired between the two conduits. The greater the distance between the shots on the second conduit, the greater the distance between the two paralleled conduits

PVC BENDING

Often, when polyvinyl chloride (PVC) is used in a conduit installation, preformed 90-degree and 45-degree fittings are purchased (**Figure 69**). Although these preformed fittings are sometimes faster and more efficient to use, learning how to effectively bend PVC can be beneficial. Preformed bends can be installed more quickly, but they are costly, and bending PVC onsite will save money. This approach is ideal if your budget does not allow for the purchase of preformed bends.

Figure 69   Preformed 45-degree and 90-degree PVC bends.

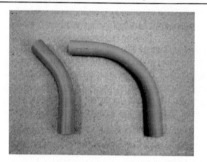

Courtesy of Alan W. Stanfield.

A few different techniques may be used to effectively bend PVC. One method relies on a "hot box" or heater (**Figure 70A**). A second method involves the use of a heat gun (**Figure 70B**). As a third option, a heating blanket can be used; some may prefer this method because of the small size and light weight of the blanket (**Figure 70C**). In the following step-by-step instructions, an electric PVC heater is used to demonstrate the process of bending PVC.

PVC BENDING

Figure 70 PVC heating devices.

A. PVC Heater

B. PVC Heat Gun

C. PVC Heating Blanket

A, C: Courtesy of Alan W. Stanfield. B. Courtesy of Ideal Industries, Inc.

Making a PVC Bend

Step 1: Warm up the electric PVC heater.

Before you begin to bend the PVC, plug in the PVC heater. It takes a few minutes for the heater to reach the temperature needed to bend the conduit. Make sure that you allow the heater to reach the highest temperature before you insert the conduit.

Step 2: Insert two PVC plugs.

This step involves inserting two PVC plugs, one in each end of the conduit (**Figure 71**). The plugs will trap the air and will help keep the conduit from collapsing while you are forming the bend. Although using plugs is not strictly necessary to bend PVC, this technique is very helpful in forming a good bend and minimizing kinking or collapsing of the conduit. Conduit plugs are available in a variety of sizes to accommodate your bending needs. If PVC plugs cannot be found in the electrical department of your retailer or wholesaler, then some plumbing departments may also carry plugs that will work.

Figure 71   Step 2: Insert two PVC plugs.

Courtesy of Alan W. Stanfield.

PVC BENDING

Step 3: Insert the conduit and heat it.

Place the PVC into the heater and slowly rotate the conduit in the heater. The heater model used in this example has rollers on each end to facilitate rotation of the conduit while you are heating it.

Rotating the conduit will ensure even heat distribution around the conduit (**Figure 72**).

Figure 72 Step 3: Insert the conduit and heat it.

Courtesy of Alan W. Stanfield.

Step 4: Heat the PVC for an appropriate amount of time.

Different conduit sizes will take different amounts of heating time before the conduit is ready to be removed from the heater and bent. As you rotate the conduit in the heater, you will notice that the PVC becomes increasingly soft. Judging the amount of time necessary for the conduit to reach the proper temperature takes experience. It is most effective to simply judge by touch. If the conduit is not heated thoroughly, it will kink (**Figure 73**). If the PVC is heated too much, it will deform or collapse. Care must be taken to avoid this problem.

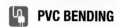

PVC BENDING

Figure 73 Step 4: Avoid kinking the conduit by heating PVC for an appropriate amount of time.

Courtesy of Alan W. Stanfield.

The following are tips on how to avoid kinks when heating and bending PVC:

- Use PVC plugs while heating.
- When forming the bend, apply equal pressure to each part of the bend.
- Form the pipe on a clean, flat, smooth surface.
- Continue to mold the pipe while it cools.
- Use a jig or other form to hold the pipe in place while it cools.

PVC BENDING

Step 5: Bend the conduit.

After the conduit has become soft, remove it from the heater and quickly bend it to the desired angle or bend. You will need to work quickly because the PVC will quickly start to harden. You may use a framing square or some other type of angle tool to ensure the proper angle is created. A preformed fitting can also be used as a template to gauge when the proper angle has been achieved. Wear leather gloves so that you can form the conduit with your hands. It is best to use a clean, flat surface to form the bend (**Figure 74**).

Figure 74   Step 5: Bend the conduit on a clean, flat surface.

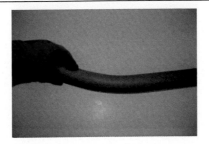

Courtesy of Alan W. Stanfield.

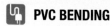

Step 6: Cool the conduit.

It is important to cool the conduit as soon as you have obtained the desired bend. PVC has memory, so the conduit will try to return or spring back into a straight position. To prevent this behavior, you must cool the conduit quickly. This can be done in a couple of ways. One option is to dip the conduit into water (**Figure 75**). Another method is to spray the conduit with a hose. The plugs can be removed during the cool-down process.

Figure 75   Step 6: Cool the conduit by dipping it in water.

Courtesy of Alan W. Stanfield.

🔌 CONDUIT FILL CALCULATIONS

Two scenarios must be considered when you are determining conduit fill. In the first scenario, all of the conductors in the conduit are the same size. In the second scenario, different sizes of conductors are installed in the conduit. You will need to refer to Chapter 9 and Informative Annex C in the 2020 *NEC®* to determine conduit fill.

Same-Size Conductors

Step 1: Refer to Tables C.1 through C.12 in Informative Annex C of the NEC® *for the type of raceway or conduit used (Tables 22 to 26).*

Start by locating the correct table for the type of conduit being used. A specific table is provided for each type of conduit; be sure to choose the correct one. Use the Fixture Wire tables, not the Compact Wire tables.

Step 2: Select the type of conductor used.

The left-hand column of each table lists the various types of conductors. Carefully choose the desired type of conductor used.

Step 3: Select the size of conduit used.

The size of the conduit is listed at the top of the table. Both metric and trades sizes are provided.

Step 4: Select the number of conductors allowed in the conduit based on the information in Steps 1 through 3.

Once you have determined the correct conduit table, conductor type, and proper size, identify where the conductor type and the conduit size intersect. This point will provide the number of conductors allowed in the conduit.

If the number of conductors installed in the conduit is already known but the size of the conduit required is not known, simply rearrange the steps to determine the proper conduit size from these same tables. Be sure to use the Fixture Wire tables, not the Compact Wire tables, unless compact wire is being used. Most conductors are classified as fixture wire.

CONDUIT FILL CALCULATIONS

Same-Size Conductors: Calculation Examples

Example 1

How many #12 THHN conductors are allowed in a ½-in. EMT?

Solution

According to *NEC*® Table C.1, 9 #12 THHN conductors are allowed.

Example 2

How many #8 THW conductors are allowed in a 1-in. RMC?

Solution

According to *NEC*® Table C.8, 8 #8 THW conductors are allowed.

Example 3

How many #10 TW conductors are allowed in a ¾-in. IMC?

Solution

According to *NEC*® Table C.4, 9 #10 TW conductors are allowed.

CONDUIT FILL CALCULATIONS

Table 22 Maximum Number of Conductors or Fixture Wires in IMC

Type	Conductor Size (AWG/kcmil)	Trade Size (in inches)									
		½	¾	1	1¼	1½	2	2½	3	3½	4
TW,	8	3	5	8	15	20	33	47	72	97	124
THHW,	10	5	9	15	27	36	59	84	130	174	224
THW,	12	7	13	21	36	49	80	113	175	234	301
THW-2	14	10	17	27	47	64	104	147	228	304	392
RHH*,	8	2	4	7	12	16	26	37	57	76	98
RHW*,	10	4	7	11	19	26	43	61	95	127	163
RHW-2*	12	5	9	14	25	34	56	79	122	163	209
	14	6	11	18	31	42	69	98	151	202	261
RHH*,	1	1	1	1	3	4	7	11	16	22	28
RHW*,	2	1	1	3	5	6	11	15	23	31	41
RHW-2*,	3	1	1	3	6	8	13	18	28	37	48
TW,	4	1	2	4	6	9	15	21	32	43	56
THHW,	6	1	3	5	9	12	20	28	43	58	75
THW,	1/0	1	1	1	3	4	6	9	14	19	24
THW-2	2/0	0	1	1	2	3	5	8	12	16	20
	3/0	0	1	1	1	3	4	6	10	13	17
	4/0	0	1	1	1	2	3	4	5	8	11	14

(continued on next page)

CONDUIT FILL CALCULATIONS

Table 22 Maximum Number of Conductors or Fixture
Wires in IMC (*continued*)

Type	Conductor Size (AWG/kcmil)	Trade Size (in inches)									
		½	¾	1	1¼	1½	2	2½	3	3½	4
RHH*,	250	0	0	1	1	1	3	4	7	9	12
RHW*,	300	0	0	1	1	1	2	4	6	8	10
RHW-2*,	350	0	0	1	1	1	2	3	5	7	9
TW,	400	0	0	0	1	1	1	3	4	6	8
THHW,	500	0	0	0	1	1	1	2	4	5	7
THW,	600	0	0	0	1	1	1	1	3	4	5
THW-2	700	0	0	0	0	1	1	1	3	4	5
	750	0	0	0	0	1	1	1	2	3	4
	800	0	0	0	0	1	1	1	2	3	4
	900	0	0	0	0	1	1	1	2	3	4
	1000	0	0	0	0	0	1	1	1	3	3
	1250	0	0	0	0	0	1	1	1	1	3
	1500	0	0	0	0	0	1	1	1	1	2
	1750	0	0	0	0	0	0	1	1	1	1
	2000	0	0	0	0	0	0	1	1	1	1
THHN,	1	1	1	2	4	5	9	13	20	27	35
THWN,	2	1	1	3	5	7	12	17	27	36	47
THWN-2	3	1	2	4	6	9	15	21	32	43	56
	4	1	3	4	8	11	17	25	38	51	66
	6	2	4	7	13	17	28	40	62	83	107
	8	3	6	10	18	24	39	56	86	115	149
	10	6	11	18	31	42	69	97	150	200	258
	12	10	17	29	49	67	109	154	238	318	410
	14	14	24	39	68	91	149	211	326	436	562
	1/0	1	1	1	3	4	8	11	17	23	29
	2/0	1	1	1	3	4	6	9	14	19	24
	3/0	0	1	1	2	3	5	7	12	16	20
	4/0	0	1	1	1	2	4	6	9	13	17

CONDUIT FILL CALCULATIONS

**Table 22   Maximum Number of Conductors or Fixture
Wires in IMC (*continued*)**

Type	Conductor Size (AWG/kcmil)	Trade Size (in inches)									
		½	¾	1	1¼	1½	2	2½	3	3½	4
THHN,	250	0	0	1	1	1	3	5	8	10	13
THWN,	300	0	0	1	1	1	3	4	7	9	12
THWN-2	350	0	0	1	1	1	2	4	6	8	10
	400	0	0	1	1	1	2	3	5	7	9
	500	0	0	0	1	1	1	3	4	6	7
	600	0	0	0	1	1	1	2	3	5	6
	700	0	0	0	1	1	1	1	3	4	5
	750	0	0	0	1	1	1	1	3	4	5
	800	0	0	0	0	1	1	1	3	4	5
	900	0	0	0	0	1	1	1	2	3	4
	1000	0	0	0	0	1	1	1	2	3	4

*Types RHH, RHW, and RHW-2 without outer covering.

Notes:

1.   This table is for concentric stranded conductors only. For compact stranded conductors, *NEC*® Table C.4(A) should be used.

2.   Two-hour, fire-rated RHH cable has ceramifiable insulation, which has much larger diameters than other RHH wires. Consult the manufacturer's conduit fill tables.

3.   This table represents an excerpt from Table C.4 of the 2020 *NEC*®. For more complete information, consult the *NEC*®.

Modified from *NEC*® Table C.4.

CONDUIT FILL CALCULATIONS

Table 23 Maximum Number of Conductors or Fixture Wires in RMC

Type	Conductor Size (AWG/kcmil)	Trade Size (in inches)											
		½	¾	1	1¼	1½	2	2½	3	3½	4	5	6
TW,	8	3	5	8	14	19	31	44	68	91	118	185	267
THHW,	10	5	9	14	25	34	56	80	123	164	212	332	480
THW,	12	7	12	19	33	45	75	107	165	221	284	446	644
THW-2	14	9	15	25	44	59	98	140	215	288	370	581	839
RHH*,	8	1	4	6	11	15	24	35	54	72	92	145	209
RHW*,	10	3	6	10	18	25	41	58	90	120	154	242	350
RHW-2*	12	5	8	13	23	32	52	75	115	154	198	311	448
	14	6	10	17	29	39	65	93	143	191	246	387	558
RHH*,	1	1	1	1	3	4	7	10	15	21	27	42	61
RHW*,	2	1	1	2	4	6	10	14	22	30	38	60	87
RHW-2*,	3	1	1	3	5	7	12	17	26	35	45	71	103
THHW,	4	1	1	3	6	8	14	20	31	41	53	83	120
THW,	6	1	3	5	8	11	18	27	41	55	71	111	160
THW-2,	1/0	0	1	1	2	3	6	8	13	18	23	36	52
TW	2/0	0	1	1	2	3	5	7	11	15	19	31	44
	3/0	0	1	1	1	2	4	6	9	13	16	26	37
	4/0	0	0	1	1	1	3	5	8	10	14	21	31
	250	0	0	1	1	1	3	4	6	8	11	17	25
	300	0	0	1	1	1	2	3	5	7	9	15	22
	350	0	0	0	1	1	1	3	5	6	8	13	19
	400	0	0	0	1	1	1	3	4	6	7	12	17
	500	0	0	0	1	1	1	2	3	5	6	10	14

CONDUIT FILL CALCULATIONS

Table 23 Maximum Number of Conductors or Fixture Wires in RMC (*continued*)

Type	Conductor Size (AWG/kcmil)	½	¾	1	1¼	1½	2	2½	3	3½	4	5	6
RHH*,	600	0	0	0	1	1	1	1	1	3	4	5	8	12
RHW*,	700	0	0	0	0	1	1	1	1	2	3	4	7	10
RHW-2*,	750	0	0	0	0	1	1	1	1	2	3	4	7	10
THHW,	800	0	0	0	0	1	1	1	1	2	3	4	6	9
THW,	900	0	0	0	0	1	1	1	1	1	3	4	6	8
THW-2,	1000	0	0	0	0	0	1	1	1	1	2	3	5	8
TW	1250	0	0	0	0	0	1	1	1	1	1	2	4	6
	1500	0	0	0	0	0	1	1	1	1	1	2	3	5
	1750	0	0	0	0	0	0	1	1	1	1	1	3	4
	2000	0	0	0	0	0	0	1	1	1	1	1	3	4
THHN,	1	1	1	1	4	5	8	12	19	25	33	51	74
THWN,	2	1	1	3	5	7	11	17	26	34	44	70	100
THWN-2	3	1	1	3	6	8	14	20	31	41	53	83	120
	4	1	2	4	7	10	16	23	36	48	62	98	141
	6	2	4	7	12	16	27	38	59	79	101	159	230
	8	3	6	9	16	22	37	53	82	109	140	221	318
	10	6	10	17	29	39	64	92	142	189	244	383	552
	12	9	16	26	46	62	102	146	225	301	387	608	877
	14	13	22	36	63	85	140	200	309	412	531	833	1202
	1/0	1	1	1	3	4	7	10	16	21	27	43	63
	2/0	0	1	1	2	3	6	8	13	18	23	36	52
	3/0	0	1	1	1	3	5	7	11	15	19	30	43
	4/0	0	1	1	1	2	4	6	9	12	16	25	36
	250	0	0	1	1	1	3	5	7	10	13	20	29
	300	0	0	1	1	1	3	4	6	8	11	17	25
	350	0	0	1	1	1	2	3	5	7	10	15	22
	400	0	0	1	1	1	2	3	5	7	8	13	20
	500	0	0	0	1	1	1	2	4	5	7	11	16

(*continued on next page*)

CONDUIT FILL CALCULATIONS

Table 23   Maximum Number of Conductors or Fixture Wires in RMC (*continued*)

Type	Conductor Size (AWG/kcmil)	Trade Size (in inches)											
		½	¾	1	1¼	1½	2	2½	3	3½	4	5	6
THHN,	600	0	0	0	1	1	1	1	3	4	6	9	13
THWN,	700	0	0	0	1	1	1	1	3	4	5	8	11
THWN-2	750	0	0	0	0	1	1	1	3	4	5	7	11
	800	0	0	0	0	1	1	1	2	3	4	7	10
	900	0	0	0	0	1	1	1	2	3	4	6	9
	1000	0	0	0	0	1	1	1	1	3	4	6	8

*Types RHH, RHW, and RHW-2 without outer covering.

Notes:

1. This table is for concentric stranded conductors only. For compact stranded conductors, *NEC*® Table C.9(A) should be used.

2. Two-hour, fire-rated RHH cable has ceramifiable insulation, which has much larger diameters than other RHH wires. Consult the manufacturer's conduit fill tables.

3. This table represents an excerpt from Table C.9 of the 2020 *NEC*®. For more complete information, consult the *NEC*®.

Modified from *NEC*® Table C.9.

CONDUIT FILL CALCULATIONS

Table 24 Maximum Number of Conductors or Fixture Wires in
Schedule 40 PVC Conduit and HDPE Conduit

Type	Conductor Size (AWG/kcmil)	Trade Size (in inches)											
		½	¾	1	1¼	1½	2	2½	3	3½	4	5	6
TW,	8	2	4	7	13	18	30	43	66	89	115	181	261
THHW,	10	4	8	13	24	32	54	77	119	160	206	325	470
THW,	12	6	11	18	32	44	72	103	160	215	277	436	631
THW-2	14	8	14	24	42	57	94	135	209	280	361	568	822
RHH*,	8	1	3	6	10	14	23	33	52	70	90	142	205
RHW*,	10	3	6	10	17	24	39	56	87	117	150	237	343
RHW-2*,	12	4	8	13	22	30	50	72	112	150	193	304	439
	14	5	9	16	28	38	63	90	139	186	240	378	546
THHW,	14	8	14	24	42	57	94	135	209	280	361	568	822
THW	8	1	3	6	10	14	23	33	52	70	90	142	205
RHH*,	1	0	1	1	3	4	7	10	15	20	26	41	60
RHW*,	2	1	1	2	4	6	10	14	22	29	37	59	85
RHW-2*,	3	1	1	3	5	7	11	16	25	34	44	69	100
THHW,	4	1	1	3	6	8	13	19	30	40	51	81	117
THW	6	1	2	4	8	11	18	26	40	53	69	109	157
THW-2,	1/0	0	1	1	2	3	6	8	13	17	22	35	51
TW	2/0	0	1	1	1	3	5	7	11	15	19	30	43
	3/0	0	1	1	1	2	4	6	9	12	16	25	36
	4/0	0	0	1	1	1	3	5	8	10	13	21	30

(continued on next page)

101

CONDUIT FILL CALCULATIONS

Table 24 Maximum Number of Conductors or Fixture Wires in Schedule 40 PVC Conduit and HDPE Conduit (*continued*)

Type	Conductor Size (AWG/kcmil)	Trade Size (in inches)											
		½	¾	1	1¼	1½	2	2½	3	3½	4	5	6
RHH*,	250	0	0	1	1	1	3	4	6	8	11	27	25
RHW*,	300	0	0	1	1	1	2	3	5	7	9	15	21
RHW-2*,	350	0	0	0	1	1	1	3	5	6	8	13	19
THHW,	400	0	0	0	1	1	1	3	4	6	7	12	17
THW,	500	0	0	0	1	1	1	2	3	5	6	10	14
THW-2,	600	0	0	0	0	1	1	1	3	4	5	8	11
TW	700	0	0	0	0	1	1	1	2	3	4	7	10
	750	0	0	0	0	1	1	1	2	3	4	6	10
	800	0	0	0	0	1	1	1	2	3	4	6	9
	900	0	0	0	0	0	1	1	1	3	3	6	8
	1000	0	0	0	0	0	1	1	1	2	3	5	7
	1250	0	0	0	0	0	1	1	1	1	2	4	6
	1500	0	0	0	0	0	1	1	1	1	1	3	5
	1750	0	0	0	0	0	0	1	1	1	1	3	4
	2000	0	0	0	0	0	0	1	1	1	1	3	4
THHN,	1	1	1	1	3	5	8	12	18	25	32	50	73
THWN,	2	1	1	3	5	7	11	16	25	33	43	68	98
THWN-2	3	1	1	3	6	8	13	19	30	40	51	81	117
	4	1	2	4	7	9	16	22	35	47	61	96	138
	6	1	4	6	11	15	26	37	57	77	99	156	225
	8	3	5	9	16	21	36	51	79	106	137	216	312
	10	5	9	15	27	37	62	89	137	184	238	374	541
	12	8	15	25	43	59	99	141	218	293	377	594	859
	14	11	21	34	60	82	135	193	299	401	517	815	1178
	1/0	1	1	1	3	4	7	10	15	21	27	42	61
	2/0	0	1	1	2	3	6	8	13	17	22	35	51
	3/0	0	1	1	1	3	5	7	11	14	18	29	42
	4/0	0	1	1	1	2	4	6	9	12	15	24	35

CONDUIT FILL CALCULATIONS

Table 24 Maximum Number of Conductors or Fixture Wires in Schedule 40 PVC Conduit and HDPE Conduit (*continued*)

Type	Conductor Size (AWG/kcmil)	Trade Size (in inches)											
		½	¾	1	1¼	1½	2	2½	3	3½	4	5	6
THHN,	250	0	0	1	1	1	3	4	7	10	12	20	28
THWN,	300	0	0	1	1	1	3	4	6	8	11	17	24
THWN-2	350	0	0	1	1	1	2	3	5	7	9	15	21
	400	0	0	0	1	1	1	3	5	6	8	13	19
	500	0	0	0	1	1	1	2	4	5	7	11	16
	600	0	0	0	1	1	1	1	3	4	5	9	13
	700	0	0	0	0	1	1	1	3	4	5	8	11
	750	0	0	0	0	1	1	1	2	3	4	7	11
	800	0	0	0	0	1	1	1	2	3	4	7	10
	900	0	0	0	0	1	1	1	2	3	4	6	9
	1000	0	0	0	0	0	1	1	1	3	3	6	8

*Types RHH, RHW, and RHW-2 without outer covering.

Notes:

1. This table is for concentric stranded conductors only. For compact stranded conductors, *NEC*® Table C.11(A) should be used.

2. Two-hour, fire-rated RHH cable has ceramifiable insulation, which has much larger diameters than other RHH wires. Consult the manufacturer's conduit fill tables.

3. This table represents an excerpt from Table C.11 of the 2020 *NEC*®. For more complete information, consult the *NEC*®.

Modified from *NEC*® Table C.11.

CONDUIT FILL CALCULATIONS

Table 25   Maximum Number of Conductors or Fixture Wires in EMT

Type	Conductor Size (AWG/kcmil)	Trade Size (in inches)									
		½	¾	1	1¼	1½	2	2½	3	3½	4
TW,	8	2	5	8	13	18	30	53	81	105	135
THHW,	10	5	8	14	24	33	55	96	145	190	243
THW,	12	6	11	19	33	45	74	129	195	255	326
THW-2	14	8	15	25	43	58	96	168	254	332	424
RHH*,	8	1	4	6	10	14	24	42	63	83	106
RHW*,	10	3	6	10	18	24	40	70	106	138	177
RHW-2*	12	4	8	13	23	31	51	90	136	177	227
	14	6	10	16	28	39	64	112	169	221	282
RHH*,	1	1	1	1	3	4	7	12	18	24	31
RHW*,	2	1	1	2	4	6	10	17	26	34	44
RHW-2*,	3	1	1	3	5	7	12	20	31	40	52
TW,	4	1	1	3	6	8	13	24	36	47	60
THW,	6	1	3	4	8	11	18	32	48	63	81
THHW,	1/0	0	1	1	2	3	6	10	16	20	26
THW-2	2/0	0	1	1	1	3	5	9	13	17	22
	3/0	0	1	1	1	2	4	7	11	15	19
	4/0	0	0	1	1	1	3	6	9	12	16

CONDUIT FILL CALCULATIONS

Table 25   Maximum Number of Conductors or Fixture Wires in EMT (*continued*)

Type	Conductor Size (AWG/kcmil)	½	¾	1	1¼	1½	2	2½	3	3½	4
RHH*,	250	0	0	1	1	1	3	5	7	10	13
RHW*,	300	0	0	1	1	1	2	4	6	8	11
RHW-2*,	350	0	0	0	1	1	1	4	6	7	10
TW,	400	0	0	0	1	1	1	3	5	7	9
THW,	500	0	0	0	1	1	1	3	4	6	7
THHW,	600	0	0	0	1	1	1	2	3	4	6
THW-2	700	0	0	0	0	1	1	1	3	4	5
	750	0	0	0	0	1	1	1	3	4	5
	800	0	0	0	0	1	1	1	3	3	5
	900	0	0	0	0	0	1	1	2	3	4
	1000	0	0	0	0	0	1	1	2	3	4
	1250	0	0	0	0	0	1	1	1	2	3
	1500	0	0	0	0	0	1	1	1	1	2
	1750	0	0	0	0	0	0	1	1	1	2
	2000	0	0	0	0	0	0	0	1	1	1
THHN,	1	1	1	1	4	5	8	15	22	29	37
THWN,	2	1	1	3	5	7	11	20	30	40	51
THWN-2	3	1	1	3	6	8	13	24	36	47	60
	4	1	2	4	7	10	16	28	43	56	71
	6	2	4	7	12	16	26	46	69	91	116
	8	3	6	9	16	22	36	64	96	126	161
	10	5	10	16	28	38	63	111	167	219	279
	12	9	16	26	45	61	101	176	266	347	443
	14	12	22	35	61	84	138	241	364	476	608
	1/0	1	1	1	3	4	7	12	19	25	32
	2/0	0	1	1	2	3	6	10	16	20	26
	3/0	0	1	1	1	3	5	8	13	17	22
	4/0	0	1	1	1	2	4	7	11	14	18

(*continued on next page*)

CONDUIT FILL CALCULATIONS

Table 25   Maximum Number of Conductors or Fixture Wires in EMT (*continued*)

Type	Conductor Size (AWG/kcmil)	Trade Size (in inches)									
		½	¾	1	1¼	1½	2	2½	3	3½	4
THHN,	250	0	0	1	1	1	3	6	9	11	15
THWN,	300	0	0	1	1	1	3	5	7	10	13
THWN-2	350	0	0	1	1	1	2	4	6	9	11
	400	0	0	0	1	1	1	4	6	8	10
	500	0	0	0	1	1	1	3	5	6	8
	600	0	0	0	1	1	1	2	4	5	7
	700	0	0	0	1	1	1	2	3	4	6
	750	0	0	0	0	1	1	1	3	4	5
	800	0	0	0	0	1	1	1	3	4	5
	900	0	0	0	0	1	1	1	3	3	4
	1000	0	0	0	0	1	1	1	2	3	4

*Types RHH, RHW, and RHW-2 without outer covering.

Note:

1. This table is for concentric stranded conductors only. For compact stranded conductors, *NEC*® Table C.1(A) should be used.

2. Two-hour, fire-rated RHH cable has ceramifiable insulation, which has much larger diameters than other RHH wires. Consult the manufacturer's conduit fill tables.

3. This table represents an excerpt from Table C.1 of the 2020 *NEC*®. For more complete information, consult the *NEC*®.

Modified from *NEC*® Table C.1.

CONDUIT FILL CALCULATIONS

Table 26   Maximum Number of Conductors or Fixture Wires in ENT

Type	Conductor Size (AWG/kcmil)	Trade Size (in inches)					
		½	¾	1	1¼	1½	2
TW,	8	2	4	7	13	18	30
THHW,	10	4	8	13	24	32	54
THW,	12	6	11	18	32	44	72
THW-2	14	8	14	24	42	57	94
RHH*,	8	1	3	6	10	14	23
RHW*,	10	3	6	10	17	24	39
RHW-2*	12	4	8	13	22	30	50
	14	5	9	16	28	38	63
RHH*,	1	0	1	1	3	4	7
RHW*,	2	1	1	2	4	6	10
RHW-2*,	3	1	1	3	5	7	11
TW,	4	1	1	3	6	8	13
THHW,	6	1	2	4	8	11	18
THW,	1/0	0	1	1	2	3	6
THW-2	2/0	0	1	1	1	3	5
	3/0	0	1	1	1	2	4
	4/0	0	0	1	1	1	3

(continued on next page)

107

CONDUIT FILL CALCULATIONS

Table 26 Maximum Number of Conductors or Fixture Wires
in ENT (*continued*)

Type	Conductor Size (AWG/kcmil)	Trade Size (in inches)					
		½	¾	1	1¼	1½	2
RHH*,	250	0	0	1	1	1	3
RHW*,	300	0	0	1	1	1	2
RHW-2*,	350	0	0	0	1	1	1
TW,	400	0	0	0	1	1	1
THHW,	500	0	0	0	1	1	1
THW,	600	0	0	0	0	1	1
THW-2	700	0	0	0	0	1	1
	750	0	0	0	0	1	1
	800	0	0	0	0	1	1
	900	0	0	0	0	0	1
	1000	0	0	0	0	0	1
	1250	0	0	0	0	0	1
	1500	0	0	0	0	0	1
	1750	0	0	0	0	0	0
	2000	0	0	0	0	0	0
THHN,	1	1	1	1	3	5	8
THWN,	2	1	1	3	5	7	11
THWN-2	3	1	1	3	6	8	13
	4	1	2	4	7	9	16
	6	1	4	6	11	15	26
	8	3	5	9	16	21	36
	10	5	9	15	27	37	62
	12	8	15	25	43	59	99
	14	11	21	34	60	82	135
	1/0	1	1	1	3	4	7
	2/0	0	1	1	2	3	6
	3/0	0	1	1	1	3	5
	4/0	0	0	1	1	2	4

108

CONDUIT FILL CALCULATIONS

Table 26 Maximum Number of Conductors or Fixture Wires in ENT (*continued*)

Type	Conductor Size (AWG/kcmil)	Trade Size (in inches)					
		½	¾	1	1¼	1½	2
THHN,	250	0	0	1	1	1	3
THWN,	300	0	0	1	1	1	3
THWN-2	350	0	0	1	1	1	2
	400	0	0	0	1	1	1
	500	0	0	0	1	1	1
	600	0	0	0	1	1	1
	700	0	0	0	0	1	1
	750	0	0	0	0	1	1
	800	0	0	0	0	1	1
	900	0	0	0	0	1	1
	1000	0	0	0	0	0	1

*Types RHH, RHW, and RHW-2 without outer covering.

Notes:

1. This table is for concentric stranded conductors only. For compact stranded conductors, *NEC*® Table C.2(A) should be used.

2. Two-hour, fire-rated RHH cable has ceramifiable insulation, which has much larger diameters than other RHH wires. Consult the manufacturer's conduit fill tables.

3. This table represents an excerpt from Table C.2 of the 2020 *NEC*®. For more complete information, consult the *NEC*®.

Modified from *NEC*® Table C.2.

CONDUIT FILL CALCULATIONS

Different-Size Conductors

If the conductors installed in a conduit are of different sizes, then an alternative method must be used. This method includes some calculating and the use of more than one table in the *NEC®*. Follow these steps:

Step 1: Refer to Chapter 9, Table 5, in the *NEC®* (**Table 27**).

Step 2: Find the size and type of conductors in Table 5 and determine the square inches of each.

Step 3: Refer to Chapter 9, Table 8, if bare conductors are present.

Step 4: Total all of the square inches of all the conductors.

Step 5: Refer to Chapter 9, Table 4, for the type and size of raceway used.

Step 6: Find the minimum size allowed based on the total area; find the allowable fill based on the number of wires. In most cases, this is a 40% fill because most raceways contain more than two wires. See Table 1 in Chapter 9 of the *NEC®*.

Step 7: Choose the appropriately sized conduit. The sum of the conductors must not be greater than the square inches allowed in the conduit.

Different-Size Conductors: Calculation Examples

Example 1

Four #6 THHN conductors and two #8 THHN conductors are to be installed in EMT. Which size should you use?

Solution

According to *NEC®* Chapter 9, Table 5:

#6 THHN	0.0507 in.^{2} × 4 conductors	=	0.2028 in.^{2}
#8 THHN	0.0366 in.^{2} × 2 conductors	=	0.0732 in.^{2}
		Total area =	0.276 in.^{2}

Using the "Over 2 Wires" column, select 1-in. EMT from *NEC®* Chapter 9, Table 4 Article 358—Electrical Metallic Tubing (EMT) (see Table 27).

CONDUIT FILL CALCULATIONS

Table 27 Dimensions of Insulated Conductors and Fixture Wires

| Type | Size (AWG or kcmil) | Approximate Area | |
		in.²	mm²
THHN, THWN, THWN-2	14	0.0097	6.258
	12	0.0133	8.581
	10	0.0211	13.61
	8	0.0366	23.61
	6	0.0507	32.71
	4	0.0824	53.16
	3	0.0973	62.77
	2	0.1158	74.71
	1	0.1562	100.8
	1/0	0.1855	119.7
	2/0	0.2223	143.4
	3/0	0.2679	172.8
	4/0	0.3237	208.8
	250	0.3970	256.1
	300	0.4608	297.3
TW, THHW, THW, THW-2	14	0.0139	8.968
	12	0.0181	11.68
	10	0.0243	15.68
	8	0.0437	28.19
	6	0.0726	46.84
	4	0.0973	62.77
	3	0.1134	73.16
	2	0.1333	86.00
	1	0.1901	122.6

(*continued on next page*)

CONDUIT FILL CALCULATIONS

Table 27 Dimensions of Insulated Conductors and
Fixture Wires (*continued*)

Type	Size (AWG or kcmil)	Approximate Area	
		in.²	mm²
TW, THHW, THW, THW-2	1/0	0.2223	143.4
	2/0	0.2624	169.3
	3/0	0.3117	201.1
	4/0	0.3718	239.9
	250	0.4596	296.5
	300	0.5281	340.7

Note: This table represents an excerpt from Chapter 9, Table 5 of the 2020 *NEC*®.
For more complete information, consult the *NEC*®.

Modified from *NEC*® Chapter 9, Table 5.

Example 2

Six #12 THW conductors and four #14 THHN conductors are to be
installed in EMT. Which size should you use?

Solution

According to *NEC*® Chapter 9, Table 5:

#12 THW	0.0181 in.² × 6 conductors	=	0.1086 in.²
#14 THHN	0.0097 in.² × 4 conductors	=	0.0388 in.²
Total area		=	0.1474 in.²

Using the "Over 2 Wires" column, select ¾-in. EMT from *NEC*®
Chapter 9, Table 4 Article 358—Electrical Metallic Tubing (EMT)
(see Table 27).

PIPE THREADING

Pipe Threading Measurements and Calculations

Sections of rigid conduit are assembled together by threading the ends and using couplings and other types of fittings. Proper care must be taken to ensure that the correct standards and calculations are used so that the fittings are securely attached to the conduit. The National Pipe Taper is a standard that is used to ensure consistent threading practice (**Table 28**).

Table 28   Excerpt from the National Pipe Taper Standard

Trade Size	Nominal Outside Diameter (OD)		Tap Drill		Threads per Inch (TPI)
	in.	mm	in.	mm	
½	0.840	21	²³⁄₃₂	18	14
¾	1.050	27	⁵⁹⁄₆₄	23	14
1	1.315	33	¹⁵⁄₃₂	29	11½
1¼	1.660	42	1½	38	11½
1½	1.900	48	1⁴⁷⁄₆₄	44	11½
2	2.375	60	1⁷⁄₃₂	56	11½
2½	2.875	73	2⅝	67	8
3	3.500	89	3¼	83	8
3½	4.000	102	3¾	95	8
4	4.500	114	4¼	108	8
5	5.563	141	5⁵⁄₃₂	134	8
6	6.625	168	6¹¹⁄₃₂	161	8

Data from the Engineering ToolBox, *NPT—National Pipe Thread Taper—ANSI B1.20.1* [http://www.engineeringtoolbox.com].

PIPE THREADING

Different sizes of conduit require a different number of threads per inch, as well as varying thread lengths. For example, a ¾-in. conduit requires 14 threads per inch at a length of ¾ in. (19 mm). Threads that are too long or too short can result in poor junctions.

Taper is a method in which the depth of the die is decreased as it cuts into the conduit, thereby changing the angle of the thread very slightly. By tapering the threads, the couplings and fittings can form a tighter joint as opposed to a straight thread with no taper.

Pipe Threading Accessories

Several accessories can assist in pipe threading (**Figure 76**):

- Cutting oil
- Oilers
- Bench yokes
- Bench chain clamps
- Tri-stand chain
- Other conduit supports

PIPE THREADING

Figure 76 Pipe threading accessories.

A. Manual Oiler

B. Cutting Oil

C. Bench Yoke

D. Bench Chain

E. Conduit Support
Courtesy of RIDGID®.

F. Stand Chain

HANDHELD MANUAL PIPE THREADERS

Handheld manual pipe threaders (**Figure 77**) are common tools that one might find in the field where a small quantity of pipe threading is required. These models are easy to transport but require much more time and work to complete the job than power pipe threaders. They generally consist of a ratcheting device with interchangeable dies for different sizes of conduit. A vise and cutting oil are needed in addition to the ratchet to complete the threading.

Figure 77   Handheld manual pipe threader.

Courtesy of RIDGID®.

HANDHELD MANUAL PIPE THREADERS

Threading Conduit Using a Handheld Manual Pipe Threader

Follow the manufacturer's instructions when threading conduit using a handheld manual pipe threader:

1. Inspect the threader before use. Replace dies or any other part that shows damage or wear. To ensure a high-quality thread, use only replacement dies available from the pipe threader's manufacturer. **Warning:** Worn dies can result in poor thread quality and high handle forces.

2. Mount the pipe firmly in a pipe vise. When threading an existing pipe, make sure it is secure and will not move. Cut the end of the pipe cleanly and squarely using a pipe cutter. **Warning:** When working on a scaffold or lift, the operator should be properly secured to prevent injury in the event of a fall.

3. Ream the end of the pipe to remove any burrs that may have been produced during the cutting of the pipe.

4. Select the correct die head for the size and type of pipe to be threaded and the thread form you require. Insert the die head into the ratchet. Slip the pipe guide at the rear of the threader over the pipe, and gently apply pressure to the front of the die head while at the same time moving the handle down to start the threader. **Warning:** Make sure the handle is clean and free from oil and grease, allowing for better control of the tool.

5. Before applying force to the handle, ensure that the ratchet pawl is engaged. **Warning:** The pawl may fail to engage if it is pushed against the side of the die head. Specific care should be taken when threading pipe that is vertical.

HANDHELD MANUAL PIPE THREADERS

6. When hand threading, your weight should be above the handle, thereby ensuring maximum leverage. If possible, do not do all the work with your arms; rather, use your weight. Do not overreach. Keep proper footing and balance to maintain better control. **Warning:** Do not slide a pipe or "cheater" over the handle to gain extra leverage, as this can result in serious injury.

7. Apply a generous quantity of thread cutting oil when threading. Use of a lubricating oil or a poor-quality thread cutting oil can result in a poor-quality cut thread, leaky joints, short die life, and high handle forces.

8. Stop threading when the end of the die is flush with the end of the pipe. At this point, the correct size has been reached to produce the proper joint. To continue beyond this point would make a straight or running thread.

9. When the thread is complete, back off the die head by reversing the ratchet mechanism and turning the die head in the opposite direction. When the die is nearly removed from the thread, maintain close control of the threader so that the threads are not damaged when you remove the die head. *Note:* Clean up any oil spill or splatter on the ground. At the end of each job, always clean your threader and store it in a clean, dry area to protect against damage.

Modified from *Manual Pipe Threader Instructions*, pages 1 to 2, with permission of RIDGID®.

HANDHELD MANUAL PIPE THREADERS

Handheld power pipe threaders (**Figure 78**) are very similar to manual models, with the exception that power threaders use a motor to assist the operator. These models are more expensive but work quickly and require much less effort to perform the threading jobs. They are a good option for jobs that require a considerable amount of threading where space is limited or when the job requires easy or frequent transport of the equipment.

Figure 78   Handheld power pipe threader.

Courtesy of RIDGID®.

HANDHELD MANUAL PIPE THREADERS

Threading Conduit Using a Handheld Power Pipe Threader

Follow the manufacturer's instructions when threading conduit using a handheld power pipe threader:

1. Push the die heads, spline end first, squarely into the power drive until the spring catches securely. *Note:* Installation can be made into either side of the power drive.

2. If possible, secure the pipe in a portable tri-stand vise or a bench vise. **Warning:** To prevent them from tipping over, long lengths of pipe should also be supported with pipe stand.

3. Make sure the oiler is properly filled with oil. Position the oiler in front of the vise.

4. Position the support arm on the pipe so that the end of the torque arm is in line with the end of the pipe. **Warning:** To avoid serious injury from losing control of the power drive, a support arm should be used when you are threading ¾-in. or larger pipe. When threading pipe less than ¾ in. in size without a support arm, hold on to the power drive firmly with one hand to exert pressure against the handle forces developed during threading.

5. Position the directional switch for the desired right- or left-hand thread. **Warning:** Change the position of the directional switch only when the motor is switched *off*.

6. Place the die head over the end of the pipe. **Warning:** To avoid serious injury, make sure the power drive is correctly positioned on the support arm. For right-hand threads, the die head will rotate clockwise (looking at the face of the die head). Forces developed by the threading torque will be in the opposite, or counterclockwise, direction.

7. Simultaneously actuate the on/off switch and exert pressure against the die head with the palm of your free hand to make sure that the thread is started. Apply plenty of thread cutting oil to the dies during threading; this will reduce the torque required to thread and improve the thread quality.

8. Keep the on/off switch depressed until the end of the pipe is even with the edge of the dies and release the switch button.

9. Back off the die head from the threaded pipe by reversing the directional switch and actuating the on/off switch. **Warning:** Firmly hold on to the power drive handle to resist handle forces that are developing while backing off the die head.

10. When the dies clear the end of the pipe, grip the handle on top of the power drive and remove the power drive and die head from the pipe.

11. Remove the support arm from the pipe and the pipe from the vise. **Warning:** To avoid injury, make sure long sections of pipe are supported at the end farthest away from the vise prior to removal.

12. Clean up any oil spills or splatter on the ground surrounding the vise and oiler.

Modified from *600 Power Drive: Operator's Manual,* pages 5 to 7, with permission of RIDGID®.

MOUNTED POWER PIPE THREADER

Mounted power pipe threading machines are the ultimate tool for quickly and accurately threading conduit. These machines are both powerful and efficient. Unfortunately, they are also expensive and difficult to navigate around a job site. This type of machine is normally found in a shop or some type of more permanent setting. **Figure 79** shows an example of a threading machine.

Figure 79 Mounted power pipe threader.

Courtesy of RIDGID®.

MOUNTED POWER PIPE THREADER

The die cutters usually have a certain range of pipe sizes that they can thread. For example, the smallest set of dies may thread ½-in. and ¾-in. pipe. Threading larger sizes of pipe may require replacing the dies with the appropriate size for larger-diameter pipe. The dies are housed in the die head. The threading machine also contains a pipe cutter and a pipe reamer; the reamer removes the burrs from the threaded pipe.

Threading Conduit Using a Maunted Power Pipe Threader

Follow the manufacturer's instructions when threading conduit using a mounted power pipe threader:

1. Put your safety glasses on and clean all the die heads of any excess metal shavings with the brush.
2. Insert the pipe into the end of the machine with 6 to 7 in. (152 to 178 mm) out the end toward the cutter and die head (**Figure 80**).

Figure 80   Secure the pipe.

MOUNTED POWER PIPE THREADER

3. Tighten both clamps to secure the pipe.
4. Move the cutter and reamer to the up position.
5. Set the die head to the appropriate setting.
6. Close the die head lever.
7. Place the machine switch in the FOR position.
8. Apply power and slowly move the die head toward the end of the pipe.
9. Apply a small amount of pressure to the die head until it grabs the pipe. The die head will feed itself to finish the threading.
10. Allow the pipe to continue threading until the end of the pipe gets within ¼ in. (6 mm) of clearing the die heads (**Figure 81**).

Figure 81 Thread the pipe.

11. While still running the machine in the FOR mode, slowly lift open the die head lever to stop threading.

12. Stop the machine and roll back the die head.

13. The pipe is now threaded.

14. Confirm that the pipe is properly threaded by using a coupling.

15. Move the die head to the up position and lower the reamer into position.

16. Apply power to the machine in the FOR mode, and slowly move the reamer into the pipe.

17. Move the reamer to the up position and lower the cutter into the cutting position.

18. Apply power to the machine, and use the cutter to cut the pipe to the desired length by constantly tightening the cutter handle.

19. Remove the pipe. Repeat the process to thread the other end.

20. Clean all excess metal shavings from the die heads and reamer.

21. Clean up any cutting oil that may have spilled.

NATURAL TRIGONOMETRIC FUNCTIONS

Angle	Sine	Cosine	Tangent	Cotangent	Secant	Cosecant	
0	.0000	1.0000	.0000		1.0000		90
1	.0175	.9998	.0175	57.2900	1.0002	57.2987	89
2	.0349	.9994	.0349	28.6363	1.0006	28.6537	88
3	.0523	.9986	.0524	19.0811	1.0014	19.1073	87
4	.0698	.9976	.0699	14.3007	1.0024	14.3356	86
5	.0872	.9962	.0875	11.4301	1.0038	11.4737	85
6	.1045	.9945	.1051	9.5144	1.0055	9.5668	84
7	.1219	.9925	.1228	8.1443	1.0075	8.2055	83
8	.1392	.9903	.1405	7.1154	1.0098	7.1853	82
9	.1564	.9877	.1584	6.3138	1.0125	6.3925	81
10	.1736	.9848	.1763	5.6713	1.0154	5.7588	80
11	.1908	.9816	.1944	5.1446	1.0187	5.2408	79
12	.2079	.9781	.2126	4.7046	1.0223	4.8097	78
13	.2250	.9744	.2309	4.3315	1.0263	4.4454	77
14	.2419	.9703	.2493	4.0108	1.0306	4.1336	76
15	.2588	.9659	.2679	3.7321	1.0353	3.8637	75
16	.2756	.9613	.2867	3.4874	1.0403	3.6280	74
17	.2924	.9563	.3057	3.2709	1.0457	3.4203	73
18	.3090	.9511	.3249	3.0777	1.0515	3.2361	72
19	.3256	.9455	.3443	2.9042	1.0576	3.0716	71
20	.3420	.9397	.3640	2.7475	1.0642	2.9238	70
21	.3584	.9336	.3839	2.6051	1.0711	2.7904	69
22	.3746	.9272	.4040	2.4751	1.0785	2.6695	68
23	.3907	.9205	.4245	2.3559	1.0864	2.5593	67
24	.4067	.9135	.4452	2.2460	1.0946	2.4586	66
25	.4226	.9063	.4663	2.1445	1.1034	2.3662	65
26	.4384	.8988	.4877	2.0503	1.1126	2.2812	64
27	.4540	.8910	.5095	1.9626	1.1223	2.2027	63
28	.4695	.8829	.5317	1.8807	1.1326	2.1301	62
29	.4848	.8746	.5543	1.8040	1.1434	2.0627	61
30	.5000	.8660	.5774	1.7321	1.1547	2.0000	60
	Cosine	**Sine**	**Cotangent**	**Tangent**	**Cosecant**	**Secant**	**Angle**
31	.5150	.8572	.6009	1.6643	1.1666	1.9416	59

126

NATURAL TRIGONOMETRIC FUNCTIONS

Angle	Sine	Cosine	Tangent	Cotangent	Secant	Cosecant	
32	.5299	.8480	.6249	1.6003	1.1792	1.8871	58
33	.5446	.8387	.6494	1.5399	1.1924	1.8361	57
34	.5592	.8290	.6745	1.4826	1.2062	1.7883	56
35	.5736	.8192	.7002	1.4281	1.2208	1.7434	55
36	.5878	.8090	.7265	1.3764	1.2361	1.7013	54
37	.6018	.7986	.7536	1.3270	1.2521	1.6616	53
38	.6157	.7880	.7813	1.2799	1.2690	1.6243	52
39	.6293	.7771	.8098	1.2349	1.2868	1.5890	51
40	.6428	.7660	.8391	1.1918	1.3054	1.5557	50
41	.6561	.7547	.8693	1.1504	1.3250	1.5243	49
42	.6691	.7431	.9004	1.1106	1.3456	1.4945	48
43	.6820	.7314	.9325	1.0724	1.3673	1.4663	47
44	.6947	.7193	.9657	1.0355	1.3902	1.4396	46
45	.7071	.7071	1.0000	1.0000	1.4142	1.4142	45
	Cosine	Sine	Cotangent	Tangent	Cosecant	Secant	Angle

Note: For angles 0 to 45, use top row and left column.

For angles 45 to 90, use bottom row and right column.

NATURAL TRIGONOMETRIC FUNCTIONS

Trigonometry

Trigonometry is the branch of mathematics dealing with the relations of sides and angles of triangles.

A **triangle** is a figure enclosed by three straight sides. The sum of the three angles is 180 degrees. All triangles have six parts: three angles and three sides opposite the angles.

Right triangles are triangles that have one angle of 90 degrees and two angles of less than 90 degrees.

To help you remember the six trigonometric functions, memorize:

"Oh Hell Another Hour of Andy"

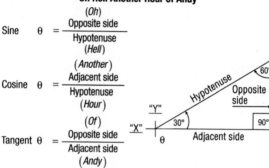

$$\text{Sine } \theta = \frac{(Oh)\ \text{Opposite side}}{\text{Hypotenuse }(Hell)}$$

$$\text{Cosine } \theta = \frac{(Another)\ \text{Adjacent side}}{\text{Hypotenuse }(Hour)}$$

$$\text{Tangent } \theta = \frac{(Of)\ \text{Opposite side}}{\text{Adjacent side }(Andy)}$$

Now, use backwards: **"Andy of Hour Another Hell Oh"**

$$\text{Cotangent } \theta = \frac{(Andy)\ \text{Adjacent side}}{\text{Opposite side }(Of)}$$

Always place the angle to be solved at the vertex (where "X" and "Y" cross)

$$\text{Secant } \theta = \frac{(Hour)\ \text{Hypotenuse}}{\text{Adjacent side }(Another)}$$

$$\text{Cosecant } \theta = \frac{(Hell)\ \text{Hypotenuse}}{\text{Opposite side }(Oh)}$$

Note:

θ = Theta = Any Angle

U.S. WEIGHTS AND MEASURES

Linear Measures

		1 Inch	= 2.540 Centimeters
12 Inches	= 1 Foot		= 3.048 Decimeters
3 Feet	= 1 Yard		= 9.144 Decimeters
5.5 Yards	= 1 Rod		= 5.029 Meters
40 Rods	= 1 Furlong		= 2.018 Hectometers
8 Furlongs	= 1 Mile		= 1.609 Kilometers

Mile Measurements

1 Statute Mile	=	5280	Feet
1 Scots Mile	=	5952	Feet
1 Irish Mile	=	6720	Feet
1 Russian Verst	=	3504	Feet
1 Italian Mile	=	4401	Feet
1 Spanish Mile	=	15084	Feet

Other Linear Measurements

1 Hand	= 4 Inches	1 Link	=	7.92	Inches
1 Span	= 9 Inches	1 Fathom	=	6	Feet
1 Chain	= 22 Yards	1 Furlong	=	10	Chains
		1 Cable	=	608	Feet

Square Measures

144	Square Inches	= 1	Square Foot
9	Square Feet	= 1	Square Yard
30¼	Square Yards	= 1	Square Rod
40	Rods	= 1	Rood
4	Roods	= 1	Acre
640	Acres	= 1	Square Mile
1	Square Mile	= 1	Section
36	Sections	= 1	Township

🔌 U.S. WEIGHTS AND MEASURES

Cubic or Solid Measures

1	Cu. Foot	=	1728 Cu. Inches
1	Cu. Yard	=	27 Cu. Feet
1	Cu. Foot	=	7.48 Gallons
1	Gallon (Water)	=	8.34 Lb.
1	Gallon (U.S.)	=	231 Cu. Inches of Water
1	Gallon (Imperial)	=	277¼ Cu. Inches of Water

Liquid Measurements

1	Pint	= 4	Gills
1	Quart	= 2	Pints
1	Gallon	= 4	Quarts
1	Firkin	= 9	Gallons (Ale or Beer)
1	Barrel	= 42	Gallons (Petroleum or Crude Oil)

Dry Measures

1	Quart	=	2 Pints
1	Peck	=	8 Quarts
1	Bushel	=	4 Pecks

Weight Measurements (Mass)

A. Avoirdupois Weight:

1 Ounce	=	16 Drams
1 Pound	=	16 Ounces
1 Hundredweight	=	100 Pounds
1 Ton	=	2000 Pounds

U.S. WEIGHTS AND MEASURES

B. Troy Weight:

1 Carat	=	3.17	Grains
1 Pennyweight	=	20	Grains
1 Ounce	=	20	Pennyweights
1 Pound	=	12	Ounces
1 Long Hundredweight	=	112	Pounds
1 Long Ton	=	20	Long Hundredweights
	=	2240	Pounds

C. Apothecaries Weight:

1 Scruple	=	20	Grains	=	1.296	Grams
1 Dram	=	3	Scruples	=	3.888	Grams
1 Ounce	=	8	Drams	=	31.1035	Grams
1 Pound	=	12	Ounces	=	373.2420	Grams

D. Kitchen Weights and Measures:

1 U.S. Pint	=	16	Fl. Ounces
1 Standard Cup	=	8	Fl. Ounces
1 Tablespoon	=	0.5	Fl. Ounces (15 Cu. Cm.)
1 Teaspoon	=	0.16	Fl. Ounces (5 Cu. Cm.)

METRIC SYSTEM

Prefixes

A. Mega	= 1000000	E. Deci	= 0.1
B. Kilo	= 1000	F. Centi	= 0.01
C. Hecto	= 100	G. Milli	= 0.001
D. Deka	= 10	H. Micro	= 0.000001

Linear Measures

(The Unit Is the Meter = 39.37 Inches)

1 Centimeter	= 10	Millimeters	= 0.3937011	In.
1 Decimeter	= 10	Centimeters	= 3.9370113	In.
1 Meter	= 10	Decimeters	= 1.0936143	Yd.
			= 3.2808429	Ft.
1 Dekameter	= 10	Meters	= 10.936143	Yd.
1 Hectometer	= 10	Dekameters	= 109.36143	Yd.
1 Kilometer	= 10	Hectometers	= 0.62137	Mile
1 Myriameter	= 10000	Meters		

Square Measures

(The Unit Is the Square Meter = 1549.9969 Sq. Inches)

1 Sq. Centimeter	= 100 Sq. Millimeters	= 0.1550	Sq. In.
1 Sq. Decimeter	= 100 Sq. Centimeters	= 15.550	Sq. In.
1 Sq. Meter	= 100 Sq. Decimeters	= 10.7639	Sq. Ft.
1 Sq. Dekameter	= 100 Sq. Meters	= 119.60	Sq. Yd.
1 Sq. Hectometer	= 100 Sq. Dekameters		
1 Sq. Kilometer	= 100 Sq. Hectometers		

(The Unit Is the "Are" = 100 Sq. Meters)

🔌 METRIC SYSTEM

1 Centiare	= 10	Milliares	=	10.7643	Sq. Ft.
1 Deciare	= 10	Centiares	=	11.96033	Sq. Yd.
1 Are	= 10	Deciares	=	119.6033	Sq. Yd.
1 Dekare	= 10	Ares	=	0.247110	Acres
1 Hektare	= 10	Dekares	=	2.471098	Acres
1 Sq. Kilometer	= 100	Hektares	=	0.38611	Sq. Mile

Cubic Measures

(The Unit Is the "Stere" = 61025.38659 Cu. In.)

1 Decistere	=	10 Centisteres	=	3.531562 Cu. Ft.
1 Stere	=	10 Decisteres	=	1.307986 Cu. Yd.
1 Dekastere	=	10 Steres	=	13.07986 Cu. Yd.

(The Unit Is the Meter = 39.37 Inches)

1 Cu. Centimeter	= 1000 Cu. Millimeters	= 0.06102 Cu. In.
1 Cu. Decimeter	= 1000 Cu. Centimeters	= 61.02374 Cu. In.
1 Cu. Meter	= 1000 Cu. Decimeters	= 35.31467 Cu. Ft.
	= 1 Stere	= 1.30795 Cu. Yd.
1 Cu. Centimeter (Water)		= 1 Gram
1000 Cu. Centimeters (Water) = 1 Liter		= 1 Kilogram
1 Cu. Meter (1000 Liters)		= 1 Metric Ton

METRIC SYSTEM

Measures of Weight

(The Unit Is the Gram = 0.035274 Ounces)

1 Milligram	=	=	0.015432	Grains
1 Centigram	= 10 Milligrams	=	0.15432	Grains
1 Decigram	= 10 Centigrams	=	1.5432	Grains
1 Gram	= 10 Decigrams	=	15.4323	Grains
1 Dekagram	= 10 Grams	=	5.6438	Drams
1 Hectogram	= 10 Dekagrams	=	3.5274	Ounces
1 Kilogram	= 10 Hectograms	=	2.2046223	Pounds
1 Myriagram	= 10 Kilograms	=	22.046223	Pounds
1 Quintal	= 10 Myriagrams	=	1.986412	Cwt.
1 Metric Ton	= 10 Quintal	= 22045.622		Pounds
1 Gram	= 0.56438 Drams			
1 Dram	= 1.77186 Grams			
	= 27.3438 Grains			
1 Metric Ton	= 2204.6223 Pounds			

Measures of Capacity

(The Unit Is the Liter = 1.0567 Liquid Quarts)

1 Centiliter	= 10 Milliliters	=	0.338	Fluid Ounces
1 Deciliter	= 10 Centiliters	=	3.38	Fluid Ounces
1 Liter	= 10 Deciliters	=	33.8	Fluid Ounces
1 Dekaliter	= 10 Liters	=	0.284	Bushel
1 Hectoliter	= 10 Dekaliters	=	2.84	Bushels
1 Kiloliter	= 10 Hectoliters	=	264.2	Gallons

Note: $\dfrac{Kilometers}{8} = 5$ *Miles* or $\dfrac{Miles}{5} \times 8$ *Kilometers*

METRIC DESIGNATORS AND TRADE SIZES

Metric Designators

12	16	21	27	35	41	53	63	78	91	103	129	155
⅜	½	¾	1	1¼	1½	2	2½	3	3½	4	5	6

Trade Sizes
U.S. Weights and Measures/Metric Equivalent Chart

	in.	ft.	yd.	mile	mm	cm	M	km
1 Inch =	1	.0833	.0278	1.578 × 10⁻⁵	25.4	**2.54**	.0254	2.54 × 10⁻³
1 Foot =	12	1	.333	1.894 × 10⁻⁴	304.8	**30.48**	.3048	3.048 × 10⁻⁴
1 Yard =	36	3	1	5.6818 × 10⁻⁴	914.4	91.44	**.9144**	9.144 × 10⁻⁴
1 Mile =	63360	5280	1760	1	1609344	160934.4	1609.344	**1609344**
1 mm =	**.03937**	.00032808	1.0936 × 10⁻³	6.2137 × 10⁻⁷	1	0.1	0.001	0.000001
1 cm =	**.3937**	.0328084	.0109361	6.2137 × 10⁻⁶	10	1	0.01	0.00001
1 m =	39.37	3.280.84	**1.093.61**	6.2137 × 10⁻⁴	1000	100	1	0.001
1 km =	39370	3280.84	1093.61	**0.62137**	1000000	100000	1000	1

In. = Inches; Ft. = Foot; Yd. = Yard; Mi. = Mile; Mm = Millimeter; Cm = Centimeter;
M = Meter; Km = Kilometer

Explanation of Scientific Notation:

Scientific notation is simply a way of expressing very large or very small numbers in a more compact format. Any number can be expressed as a number between 1 and 10, multiplied by a power of 10 (which indicates the correct position of the decimal point in the original number). Numbers greater than 10 have positive powers of 10, and numbers less than 1 have negative powers of 10.

Example: $186,000 = 1.86 \times 10^5$ $0.000524 = 5.24 \times 10^{-4}$

METRIC DESIGNATORS AND TRADE SIZES

Useful Conversions/Equivalents

1 BTU	Raises 1 lb. of water 1°F
1 Gram Calorie	Raises 1 gram of water 1°C
1 Circular Mil	Equals 0.7854 sq. mil
1 Sq. Mil	Equals 1.27 cir. mils
1 Mil	Equals 0.001 in.

To determine circular mil (CM) of a conductor:

Round Conductor ....CM = (Diameter in mils)^2

Bus Bar................................CM = $\dfrac{\text{Width (mils)} \times \text{Thickness (mils)}}{0.7854}$

Notes: 1 millimeter = 39.37 mils   1 cir. millimeter = 1550 cir. mils

1 sq. millimeter = 1974 cir. mils

DECIMAL EQUIVALENTS

	Fraction				Decimal		Fraction				Decimal
1/64					.0156	33/64					.5156
2/64	1/32				.0313	34/64	17/32				.5313
3/64					.0469	35/64					.5469
4/64	2/32	1/16			.0625	36/64	18/32	9/16			.5625
5/64					.0781	37/64					.5781
6/64	3/32				.0938	38/64	19/32				.5938
7/64					.1094	39/64					.6094
8/64	4/32	2/16	1/8		.125	40/64	20/32	10/16	5/8		.625
9/64					.1406	41/64					.6406
10/64	5/32				.1563	42/64	21/32				.6563
11/64					.1719	43/64					.6719
12/64	6/32	3/16			.1875	44/64	22/32	11/16			.6875
13/64					.2031	45/64					.7031
14/64	7/32				.2188	46/64	23/32				.7188
15/64					.2344	47/64					.7344
16/64	8/32	4/16	2/8	1/4	.25	48/64	24/32	12/16	6/8	3/4	.75
17/64					.2656	49/64					.7656
18/64	9/32				.2813	50/64	25/32				.7813
19/64					.2969	51/64					.7969
20/64	10/32	5/16			.3125	52/64	26/32	13/16			.8125
21/64					.3281	53/64					.8281
22/64	11/32				.3438	54/64	27/32				.8438
23/64					.3594	55/64					.8594
24/64	12/32	6/16	3/8		.375	56/64	28/32	14/16	7/8		.875
25/64					.3906	57/64					.8906

DECIMAL EQUIVALENTS

	Fraction				Decimal		Fraction				Decimal
26/64	13/32				.4063	58/64	29/32				.9063
27/64					.4219	59/64					.9219
28/64	14/32	7/16			.4375	60/64	30/32	15/16			.9375
29/64					.4531	61/64					.9531
30/64	15/32				.4688	62/64	31/32				.9688
31/64					.4844	63/64					.9844
32/64	16/32	8/16	4/8	2/4	.5,	64/64	32/32	16/16	8/8	4/4	1.000

Decimals are rounded to the nearest 10000th.

TWO-WAY CONVERSION TABLE

To convert from the unit of measure in Column B to the unit of measure in Column C, multiply the number of units in Column B by the multiplier in Column A. To convert from Column C to B, use the multiplier in Column D.

Example: To convert 1000 BTUs to Calories, find the "BTU–Calorie" combination in Columns B and C. "BTU" is in Column B and "Calorie" is in Column C, so we are converting from B to C. Therefore, we use the Column A multiplier: 1000 BTUs × 251.996 = 251996 Calories.

To convert 251,996 Calories to BTUs, use the same "BTU–Calorie" combination. But this time you are converting from C to B. Therefore, use the Column D multiplier: 251,996 Calories × .0039683 = 1000 BTUs.

$$A \times B = C \qquad \& \qquad D \times C = B$$

To convert from B to C, multiply B × A:			To convert from C to B, multiply C × D:
A	**B**	**C**	**D**
43,560	Acre	Sq. foot	2.2956×10^{-5}
1.5625×10^{-3}	Acre	Sq. mile	640
6.4516	Ampere per sq. cm.	Ampere per sq. in.	.155003
1.256637	Ampere (turn)	Gilberts	0.79578
33.89854	Atmosphere	Foot of H_2O	0.029499
29.92125	Atmosphere	Inch of Hg	0.033421
14.69595	Atmosphere	Pound force/ sq. in.	0.06804
251.996	BTU	Calorie	3.96832×10^{-3}
778.169	BTU	Foot pound force	1.28507×10^{-3}
3.93015×10^{-4}	BTU	Horsepower-hour	2544.43
1055.056	BTU	Joule	9.47817×10^{-4}
2.9307×10^{-4}	BTU	Kilowatt-hour	3412.14
3.93015×10^{-4}	BTU/hour	Horsepower	2544.43
2.93071×10^{-4}	BTU/hour	Kilowatt	3412.1412
0.293071	BTU/hour	Watt	3.41214
4.19993	BTU/minute	Calorie/second	0.23809

(*continued on next page*)

TWO-WAY CONVERSION TABLE

To convert from B to C, multiply B × A:			To convert from C to B, multiply C × D:
A	**B**	**C**	**D**
0.0235809	BTU/minute	Horsepower	42.4072
17.5843	BTU/minute	Watt	0.0568
4.1868	Calorie	Joule	.238846
0.0328084	Centimeter	Foot	30.48
0.3937	Centimeter	Inch	2.54
0.00001	Centimeter	Kilometer	100,000
0.01	Centimeter	Meter	100
6.2137×10^{-6}	Centimeter	Mile	160,934.4
10	Centimeter	Millimeter	0.1
0.010936	Centimeter	Yard	91.44
7.85398×10^{-7}	Circular mil	Sq. inch	1.273239×10^{6}
0.000507	Circular mil	Sq. millimeter	1973.525
0.06102374	Cubic centimeter	Cubic inch	16.387065
0.028317	Cubic foot	Cubic meter	35.31467
1.0197×10^{-3}	Dyne	Gram force	980.665
1×10^{-5}	Dyne	Newton	100,000
1	Dyne centimeter	Erg	1
7.376×10^{-8}	Erg	Foot pound force	1.355818×10^{7}
2.777×10^{-14}	Erg	Kilowatt-hour	3.6×10^{13}
1.0×10^{-7}	Erg/second	Watt	1.0×10^{7}
12	Foot	Inch	0.0833
3.048×10^{-4}	Foot	Kilometer	3280.84
0.3048	Foot	Meter	3.28084
1.894×10^{-4}	Foot	Mile	5280
304.8	Foot	Millimeter	0.00328
0.333	Foot	Yard	3
10.7639	Foot candle	Lux	0.0929
0.882671	Foot of H_2O	Inch of Hg	1.13292
5.0505×10^{-7}	Foot pound force	Horsepower-hour	1.98×10^{6}
1.35582	Foot pound force	Joule	0.737562
3.76616×10^{-7}	Foot pound force	Kilowatt-hour	2.655223×10^{6}
3.76616×10^{-4}	Foot pound force	Watt-hour	2655.22

TWO-WAY CONVERSION TABLE

To convert from B to C, multiply B × A:			To convert from C to B, multiply C × D:
A	**B**	**C**	**D**
3.76616×10^{-7}	Foot pound force/ hour	Kilowatt	2.6552×10^{6}
3.0303×10^{-5}	Foot pound force/ minute	Horsepower	33000
2.2597×10^{-5}	Foot pound force/ minute	Kilowatt	44253.7
0.022597	Foot pound force/ minute	Watt	44.2537
1.81818×10^{-3}	Foot pound force/ second	Horsepower	550
1.355818×10^{-3}	Foot pound force/ second	Kilowatt	737.562
0.7457	Horsepower	Kilowatt	1.34102
745.7	Horsepower	Watt	0.00134
.0022046	Gram	Pound mass	453.592
2.54×10^{-5}	Inch	Kilometer	39370
0.0254	Inch	Meter	39.37
1.578×10^{-5}	Inch	Mile	63360
25.4	Inch	Millimeter	0.03937
0.0278	Inch	Yard	36
0.07355	Inch of H_2O	Inch of Hg	13.5951
2.7777×10^{-7}	Joule	Kilowatt-hour	3.6×10^{6}
2.7777×10^{-4}	Joule	Watt hour	3600
1	Joule	Watt second	1
1000	Kilometer	Meter	0.001
0.62137	Kilometer	Mile	1.609344
1,000,000	Kilometer	Millimeter	0.000001
1093.61	Kilometer	Yard	9.144×10^{-4}
0.000621	Meter	Mile	1609.344
1000	Meter	Millimeter	0.001
1.0936	Meter	Yard	0.9144

(continued on next page)

TWO-WAY CONVERSION TABLE

To convert from B to C, multiply B × A:			To convert from C to B, multiply C × D:
A	**B**	**C**	**D**
1,609,344	Mile	Millimeter	6.2137×10^{-7}
1760	Mile	Yard	5.681×10^{-4}
1.0936×10^{-3}	Millimeter	Yard	914.4
0.224809	Newton	Pound force	4.44822
0.03108	Pound	Slug	32.174
0.0005	Pound	Ton (short)	2000
0.155	Sq. centimeter	Sq. inch	6.4516
0.092903	Sq. foot	Sq. meter	10.76391
0.386102	Sq. kilometer	Sq. mile	2.589988

CENTIGRADE AND FAHRENHEIT THERMOMETER SCALES

DEG-C	DEG-F	DEG-C	DEG-F	DEG-C	DEG-F	DEG-C	DEG-F
0	32						
1	33.8	26	78.8	51	123.8	76	168.8
2	35.6	27	80.6	52	125.6	77	170.6
3	37.4	28	82.4	53	127.4	78	172.4
4	39.2	29	84.2	54	129.2	79	174.2
5	41	30	86	55	131	80	176
6	42.8	31	87.8	56	132.8	81	177.8
7	44.6	32	89.6	57	134.6	82	179.6
8	46.4	33	91.4	58	136.4	83	181.4
9	48.2	34	93.2	59	138.2	84	183.2
10	50	35	95	60	140	85	185
11	51.8	36	96.8	61	141.8	86	186.8
12	53.6	37	98.6	62	143.6	87	188.6
13	55.4	38	100.4	63	145.4	88	190.4
14	57.2	39	102.2	64	147.2	89	192.2
15	59	40	104	65	149	90	194
16	60.8	41	105.8	66	150.8	91	195.8
17	62.6	42	107.6	67	152.6	92	197.6
18	64.4	43	109.4	68	154.4	93	199.4
19	66.2	44	111.2	69	156.2	94	201.2
20	68	45	113	70	158	95	203
21	69.8	46	114.8	71	159.8	96	204.8
22	71.6	47	116.6	72	161.6	97	206.6
23	73.4	48	118.4	73	163.4	98	208.4
24	75.2	49	120.2	74	165.2	99	210.2
25	77	50	122	75	167	100	212

1. Temperature $C° = \frac{5}{9} \times$ (Temperature $F° - 32$).

2. Temperature $F° = (\frac{9}{5} \times$ Temperature $C°) + 32$.

3. Ambient temperature is the temperature of the surrounding cooling medium.

4. Rated temperature rise is the permissible rise in temperature above ambient when operating under load.

 **FRACTIONS**

Definitions

A. A *fraction* is a quantity less than a unit.

B. A *numerator* is the term of a fraction indicating how many of the parts of a unit are to be taken. In a common fraction, it appears above or to the left of the line.

C. A *denominator* is the term of a fraction indicating the number of equal parts into which the unit is divided. In a common fraction, it appears below or to the right of the line.

D. Examples:

1. $\dfrac{1}{2}$ $=$ $\dfrac{\text{Numerator}}{\text{Denominator}}$ = Fraction

2. Numerator $\longrightarrow$ ½ $\longleftarrow$ Denominator

To Add or Subtract

To Solve: ½ – ⅔ + ¾ – ⅚ + ⁷⁄₁₂ = ?

A. Determine the lowest common denominator that each of the denominators 2, 3, 4, 6, and 12 will divide into an even number of times.

The lowest common denominator is 12.

B. Work one fraction at a time using the formula:

$$\dfrac{\text{Common Denominator}}{\text{Denominator of Fraction}} \times \text{Numerator of Fraction}$$

1. ¹²⁄₂ × 1 = 6 × 1 = 6   ½ becomes ⁶⁄₁₂
2. ¹²⁄₃ × 2 = 4 × 2 = 8   ⅔ becomes ⁸⁄₁₂
3. ¹²⁄₄ × 3 = 3 × 3 = 9   ¾ becomes ⁹⁄₁₂
4. ¹²⁄₆ × 5 = 2 × 5 = 10   ⅚ becomes ¹⁰⁄₁₂
5. ⁷⁄₁₂ remains ⁷⁄₁₂

FRACTIONS

C. We can now convert the problem from its original form to its new form using 12 as the common denominator.

$$½ - ⅔ + ¾ - ⅚ + \frac{7}{12} = \text{Original form}$$

$$\frac{6 - 8 + 9 - 10 + 7}{12} = \text{Present form}$$

$$\frac{4}{12} = \frac{1}{3} \text{ Reduced to lowest form}$$

D. To convert fractions to decimal form, simply divide the numerator of the fraction by the denominator of the fraction.

Example: $\frac{1}{3} = 1$ Divided by 3 = 0.333

To Multiply

A. The numerator of fraction #1 times the numerator of fraction #2 is equal to the numerator of the product.

B. The denominator of fraction #1 times the denominator of fraction #2 is equal to the denominator of the product.

C. Example:

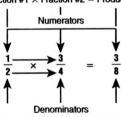

Fraction #1 × Fraction #2 = Product

Numerators

$$\frac{1}{2} \times \frac{3}{4} = \frac{3}{8}$$

Denominators

Note: To change ⅔ to decimal form, divide 3 by 8 = .375.

145

FRACTIONS

To Divide

A. The numerator of fraction #1 times the denominator of fraction #2 is equal to the numerator of the quotient.

B. The denominator of fraction #1 times the numerator of fraction #2 is equal to the denominator of the quotient.

C. Example: $\dfrac{1}{2} \div \dfrac{3}{4}$

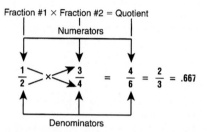

D. An a Alternate method for dividing by a fraction is to multiply by the reciprocal of the divisor (the second fraction in a division problem).

E. Example: $\dfrac{1}{2} \div \dfrac{3}{4}$

The reciprocal of $\dfrac{3}{4}$ is $\dfrac{4}{3}$

so, $\dfrac{1}{2} \div \dfrac{3}{4} = \dfrac{1}{2} \times \dfrac{4}{3} = \dfrac{4}{6} = \dfrac{2}{3} = .667$

🔌 EQUATIONS

The word "equation" means equal or the same as.

Example: $2 \times 10 = 4 \times 5$

$$20 = 20$$

Rules:

A. The same number may be added to both sides of an equation without changing its values.

Example: $(2 \times 10) + 3 = (4 \times 5) + 3$

$$23 = 23$$

B. The same number may be subtracted from both sides of an equation without changing its values.

Example: $(2 \times 10) - 3 = (4 \times 5) - 3$

$$17 = 17$$

C. Both sides of an equation may be divided by the same number without changing its values.

Example: $\dfrac{2 \times 10}{20} = \dfrac{4 \times 5}{20}$

$$1 = 1$$

D. Both sides of an equation may be multiplied by the same number without changing its values.

Example: $3 \times (2 \times 10) = 3 \times (4 \times 5)$

$$60 = 60$$

E. Transposition is the process of moving a quantity from one side of an equation to the other side of an equation by changing its sign of operation.

1. A term may be transposed if its sign is changed from plus (+) to minus (−), or from minus (−) to plus (+).

Example: $X + 5 = 25$

$$X + 5 - 5 = 25 - 5$$

$$X = 20$$

⚡ EQUATIONS

2. A multiplier may be removed from one side of an equation by making it a divisor on the other side, or a divisor may be removed from one side of an equation by making it a multiplier on the other side.

Example: Multiplier from one side of equation (4) becomes divisor on the other side.

$$4X = 40 \text{ becomes } X = \frac{40}{4} = 10$$

Example: Divisor from one side of equation becomes multiplier on the other side.

$$\frac{X}{4} = 10 \text{ becomes } X = 10 \times 4$$

Signs:

A. Addition of numbers with *different* signs:

1. Rule: Use the sign of the larger and subtract.

Example: $\dfrac{\begin{array}{r} +3 \\ +-2 \end{array}}{+1} \qquad \dfrac{\begin{array}{r} -2 \\ ++3 \end{array}}{+1}$

B. Addition of numbers with the *same* signs:

2. Rule: Use the common sign and add.

Example: $\dfrac{\begin{array}{r} +3 \\ ++2 \end{array}}{+5} \qquad \dfrac{\begin{array}{r} -3 \\ +-2 \end{array}}{-5}$

C. Subtraction of numbers with *different* signs:

3. Rule: Change the sign of the subtrahend (the second number in a subtraction problem) and proceed as in addition.

Example: $\dfrac{\begin{array}{r} +3 \\ --2 \end{array}}{+5} = \dfrac{\begin{array}{r} +3 \\ ++2 \end{array}}{+5} \qquad \dfrac{\begin{array}{r} -2 \\ -+3 \end{array}}{-5} = \dfrac{\begin{array}{r} -2 \\ +-3 \end{array}}{-5}$

EQUATIONS

D. Subtraction of numbers with the *same* signs:

    4. Rule: Change the sign of the subtrahend (the second number in a subtraction problem) and proceed as in addition.

$$\text{Example:} \quad \frac{\begin{array}{r} +3 \\ -+3 \end{array}}{+1} = \frac{\begin{array}{r} +3 \\ +-2 \end{array}}{} \quad \frac{\begin{array}{r} -3 \\ --2 \end{array}}{} = \frac{\begin{array}{r} -3 \\ ++2 \end{array}}{-1}$$

E. Multiplication:

    5. Rule: The product of any two numbers having *like* signs is *positive*. The product of any two numbers having *unlike* signs is *negative*.

    Example:
$$(+3) \times (-2) = -6$$
$$(-3) \times (+2) = -6$$
$$(+3) \times (+2) = +6$$
$$(-3) \times (-2) = +6$$

F. Division:

    6. Rule: If the divisor and the dividend have *like* signs, the sign of the quotient is *positive*. If the divisor and dividend have *unlike* signs, the sign of the quotient is *negative*.

    Example:

$$\frac{+6}{-2} = -3 \qquad \frac{+6}{+2} = +3$$

$$\frac{-6}{+2} = -3 \qquad \frac{-6}{-2} = +3$$

⚡ ELECTRICAL SAFETY DEFINITIONS

(Courtesy of Littelfuse® POWR-GARD® Products)

Approach boundaries—Protection boundaries established to protect personnel from shock.

Arc-blast—A pressure wave created by the heating, melting, vaporization, and expansion of conducting material and surrounding gases of air.

Arc-flash—The sudden release of heat energy and intense light at the point of an arc. It can be considered a short circuit through the air, usually created by accidental contact between live conductors.

Arc gap—The distance between energized conductors or between energized conductors and ground. Shorter arc gaps result in less energy being expended in the arc, whereas longer gaps reduce arc current. For 600 volts and below, arc gaps of 1.25 in. (32 mm) typically produce the maximum incident energy.

Calorie—The amount of heat needed to raise the temperature of one gram of water by one degree Celsius; 1 cal/cm^2 is equivalent to the exposure of the tip of a finger to a cigarette lighter for one second.

Distance to arc—Refers to the distance from the receiving surface to the arc center. The value used for most calculations is typically 18 in. (457 mm).

Electrically safe work condition—Condition where the equipment and/or circuit components have been disconnected from electrical energy sources, locked/tagged out, and tested to verify all sources of power are removed.

Exposed live parts—An energized conductor or part that is capable of being inadvertently touched or approached (nearer than a safe distance) by a person. It is applicable to parts that are not in an electrically safe work condition, suitably grounded, isolated, or insulated.

Reprinted with permission from Littelfuse®; www.littelfuse.com; 1-800-TEC-FUSE.

For more information, refer to *NFPA 70E®, Standard for Electrical Safety in the Workplace.*

⚡ ELECTRICAL SAFETY DEFINITIONS

Flame resistant (FR)—A term referring to fabric and its ability to limit the ignition or burning of the garment. It can be a specific characteristic of the material or a treatment applied to a material.

Flash hazard analysis—A study that analyzes potential exposure to arc-flash hazards. The outcome of the study establishes incident energy levels, hazard risk categories, flash protection boundaries, and required personal protective equipment (PPE). It also helps define safe work practices.

Flash protection boundary—A protection boundary established to protect personnel from arc-flash hazards. The flash protection boundary is the distance at which an unprotected worker can receive a second-degree burn to bare skin.

Flash suit—A term referring to a complete FR-rated PPE system that would cover a person's body, excluding the hands and feet. Included would be pants, shirt/jacket, and flash hood with a built-in face shield.

Hazard risk category—A classification of risks (from 1 to 4) defined by *NFPA 70E*. Each category requires PPE and is related to incident energy levels.

Incident energy—The amount of thermal energy impressed on a surface generated during an electrical arc at a certain distance from the arc. It is typically measured in cal/cm^2.

PPE—An acronym for personal protective equipment. It can include clothing, tools, and equipment.

Qualified person—A person who is trained and knowledgeable on the construction and operation of the equipment and can recognize and avoid electrical hazards that may be encountered.

Shock—A trauma subjected to the body by electrical current. When personnel come in contact with energized conductors, it can result in current flowing through the body, often causing serious injury or death.

Unqualified person—A person who does not possess all the needed skills and knowledge or has not been trained for a particular task.

Reprinted with permission from Littelfuse®; www.littelfuse.com; 1-800-TEC-FUSE

For more information, refer to *NFPA 70E®, Standard for Electrical Safety in the Workplace.*

ELECTRICAL SAFETY DEFINITIONS

1. Whenever possible, de-energize the equipment prior to performing any work.
2. Verify you are "qualified" and properly trained to perform the required task.
3. Identify the equipment and verify you have a clear understanding of and have been trained on how the equipment operates.
4. Provide justification of why the work must be performed in an "energized" condition (if applicable).
5. Identify what safe work practices will be used to ensure safety.
6. Determine if a hazard analysis has been performed to identify all hazards (shock, arc-flash, etc.).
7. Identify protection boundaries for shock (limited and restricted approach) and arc-flash (flash protection boundary).
8. Identify the required PPE for the task to be performed based on the hazard risk category (HRC) and available incident energy (cal/cm^2).
9. Provide barriers or other means to prevent access to the work area by "unqualified" workers.
10. Perform a job briefing and identify job- or task-specific hazards.
11. Obtain written management approval to perform the work in an "energized" condition (where applicable).

Reprinted with permission from Littelfuse®; www.littelfuse.com; 1-800-TEC-FUSE
For more information, refer to *NFPA 70E®, Standard for Electrical Safety in the Workplace.*

ELECTRICAL SAFETY: LOCKOUT–TAGOUT PROCEDURES

(*Courtesy of* Littelfuse® POWR-GARD® Products)

The Occupational Safety and Health Administration (OSHA) requires that energy sources to machines or pieces of equipment must be turned off and disconnected, isolating them from the energy source. The isolating or disconnecting means must be either locked or tagged with a warning label. While lockout is the more reliable and preferred method, OSHA accepts tagout to be a suitable replacement in limited situations. *NFPA 70E* Article 120 contains detailed instructions for lockout/tagout and placing equipment in an electrically safe work condition.

Application of Lockout–Tagout Devices

1. Make necessary preparations for shutdown.
2. Shut down the machine or equipment.
3. Turn off (open) the energy isolating device (fuse/circuit breaker).
4. Apply the lockout or tagout device.
5. Render safe all stored or residual energy.
6. Verify the isolation and de-energization of the machine or equipment.

Removal of Lockout–Tagout Devices

1. Inspect the work area to ensure that nonessential items have been removed and that machine or equipment components are intact and capable of operating properly. Especially look for tools or pieces of conductors that may not have been removed.
2. Check the area around the machine or equipment to ensure that all employees have been safely positioned or removed.
3. Make sure that only the employees who attached the locks or tags are the ones who are removing them.
4. After removing locks or tags, notify affected employees before starting equipment or machines.

Note: For specific lockout–tagout procedures, refer to OSHA and *NFPA 70E*.

Reprinted with permission from Littelfuse®; www.littelfuse.com; 1-800-TEC-FUSE.

ELECTRICAL SAFETY: HOW TO READ A WARNING LABEL

(*Courtesy of* Littelfuse® POWR-GARD® Products)

The amount of heat energy (cal/cm²) at the distance shown. The incident energy determines the hazard risk category.

The distance from exposed energized parts at which a second-degree burn can occur to unprotected skin.

The *NFPA 70E* established hazard risk category (see *NFPA 70E* Table 130.7(c)(11) for explanation) based on incident energy.

⚠ **WARNING**

Arc-Flash and Shock Hazard
Appropriate PPE Required

ARC-FLASH PROTECTION BOUNDARY AND REQUIRED PPE

Flash Hazard Boundary <u>41 inches</u>

Hazard Risk Category <u>2</u>

Incident Energy at 18" (cal/cm²) <u>4.05 (cal/cm²)</u>

Glove Class <u>00</u>

Required PPE FR Shirt and Pants + Safety Glasses + Hard Hat + Leather Gloves and Shoes + Ear Plugs + Face Shield

SHOCK HAZARD PROTECTION BOUNDARIES

Shock Hazard <u>480 VAC</u>

Restricted <u>12 inches</u>

Limited <u>42 inches</u>

Prohibited <u>1 inch</u>

Equipment ID: <u>Panel L-10</u>   Assessment Date: <u>9/18/09</u>

Equipment voltage determining the shock approach boundaries.

Name or ID of specific electrical equipment for which this label is produced.

Required PPE (personal protective equipment) based on the incident energy and hazard risk category.

Required glove class to protect against voltage and shock hazard.

Reprinted with permission from Littelfuse®; www.littelfuse.com; 1-800-TEC-FUSE.

🔌 FIRST AID

In addition to consistently wearing PPE, the best protection from the consequences of severe injury is the knowledge of first-aid practices and the ability to act in an emergency situation. The Emergency Care and Safety Institute is an educational organization created for the purpose of delivering the highest-quality training to laypeople and professionals in the areas of first aid, adult cardiopulmonary resuscitation (CPR), use of an automated external defibrillator (AED), blood-borne pathogens, and related safety and health areas. The content of the training materials used by the Emergency Care and Safety Institute is approved by the American Academy of Orthopaedic Surgeons (AAOS) and the American College of Emergency Physicians (ACEP), two of the most respected names in injury, illness, and emergency medical care.

Visit www.ECSInstitute.org for more information.

Scene Survey

When approaching the scene of an emergency, take a few seconds to perform a scene survey to assess the following:

- Danger to you, the victim(s), or bystanders. Scan the area for immediate dangers. If the scene is unsafe, make it safe. If you are unable to do so, do not enter.

- Cause of injury or nature of illness. This helps to identify what is wrong.

- Number of victims. Determine how many people are involved. There may be more than one person, so look around and ask about others.

If there are two or more victims, first check those who are not moving or talking. These are the individuals who may need your help first.

🔌 FIRST AID

How to Call for Help

To receive emergency assistance of every kind in most communities, simply call 9-1-1. At some government installations and industrial sites, an additional system may apply. This should be an element of a job briefing. In any case, be prepared to tell the emergency medical services (EMS) dispatcher the following:

- Your name and phone number.
- Exact location or address of emergency.
- What happened.
- Number of people needing help and any special conditions.
- Victim's condition and what is being done for the victim.

Do not hang up until the dispatcher hangs up—the EMS dispatcher may be able to tell you how to care for the victim until the ambulance arrives.

Bleeding

1. Cover the entire wound with a clean, dry cloth or sterile dressing.
2. Apply direct pressure to the wound. If the bleeding does not slow or stop, apply more dressings and pressure.
3. For a shallow wound, wash it with soap and water and flush with forceful, running water. For a deep wound, do not use soap and water under pressure. Deep wounds require cleaning by a medically trained person. Cover large, gaping wounds with sterile gauze pads.
4. When the bleeding stops or subsides, secure gauze or cloth snugly with a bandage. For a shallow wound, antibiotic ointment can be applied before the dressings. Do not use antibiotic ointment on a deep wound.

⚡ FIRST AID

Protect yourself against diseases carried by blood by wearing disposable medical exam gloves, using several layers of cloth or gauze pads, using waterproof material such as plastic, or having the victim apply pressure using his or her own hand.

Ingested Poisons

Recognizing Ingested Poisoning

The signs of ingested poisoning include the following:

- Abdominal pain and cramping.
- Nausea or vomiting.
- Diarrhea.
- Burns, odors, or stains around and in the mouth.
- Drowsiness or unresponsiveness.
- Poison container nearby.

Care for Ingested Poisoning

1. Remove any objects from the victim's mouth *only* if you can see them.

   - Try to determine what poison was swallowed and how much. Also make note of the age and size of the victim.

   - If the patient is responsive, call the Poison Control Center immediately for instructions: 800-222-1222.

2. If the patient is unresponsive, check breathing and treat accordingly. Call 9-1-1.

Do not give water or milk to dilute poisons unless instructed to do so by the Poison Control Center.

🔌 FIRST AID

Shock

Recognizing Shock

The signs of shock include the following:

- Altered mental status (agitation, anxiety, restlessness, and confusion).
- Pale, cold, and clammy skin, lips, and nail beds.
- Nausea and vomiting.
- Rapid breathing.
- Unresponsiveness (when shock is severe).

Care for Shock

Even if there are no signs of shock, you should still treat seriously injured or suddenly ill victims for shock.

1. Place the victim on his or her back.
2. Raise the legs 6 to 12 in. (if spinal injury is not suspected). Raising the legs allows the blood to drain from the legs back to the heart.
3. Maintain normal body temperature.
4. Call 9-1-1.

Burns

Care for Burns

1. Stop the burning! Use water or smother flames.
2. Cool the burn. Apply cool water or cool, wet cloths until pain is gone (usually within 10–40 minutes).

FIRST AID

3. Apply aloe-vera gel on first-degree burns (skin turns red). Apply antibiotic ointment on second-degree burns (skin blisters). Apply non-stick dressing on second- and third-degree burns (full thickness; skin dies).

Electrical Burns

- Check the scene for electrical hazards.
- Check breathing and provide care as needed.

Frostbite

Recognizing Frostbite

The signs of frostbite include the following:

- White, waxy-looking skin.
- Skin feels cold and numb (pain at first, followed by numbness).
- Blisters, which may appear after re-warming.

Care for Frostbite

1. Move the victim to a warm place.
2. Remove wet/cold clothing and jewelry from the injured part.
3. Seek medical care.

⚡ FIRST AID

Hypothermia

Recognizing Hypothermia

The signs of hypothermia include the following:

- Uncontrollable shivering.
- Confusion, sluggishness.
- Cold skin (even under clothing).

Care for Hypothermia

1. Get the victim out of the cold.
2. Prevent heat loss by:
   - replacing wet clothing with dry clothing.
   - covering the victim's head.
   - placing insulation (such as blankets, towels, coats) beneath and over the victim.
3. Have the victim lie down.
4. If the victim is alert and able to swallow, give him or her warm, sugary beverages.
5. Seek medical care for severe hypothermia (rigid muscles, cold skin on abdomen, confusion, or lethargy).

⚡ FIRST AID

Heat Cramps

Recognizing Heat Cramps

The signs of heat cramps include the following:

- Painful muscle spasms that occur suddenly, often after physical exertion.

Care for Heat Cramps

1. Have the victim stop activity and rest in a cool area.
2. Stretch the cramped muscle.
3. If the victim is responsive and not nauseated, provide water or a commercial sports drink (such as Gatorade® or Powerade®).

Heat Exhaustion

Recognizing Heat Exhaustion

The signs of heat exhaustion can include the following:

- Heavy sweating.
- Severe thirst.
- Weakness.
- Headache.
- Nausea and vomiting.

Care for Heat Exhaustion

1. Have the victim stop activity and rest in a cool area.
2. Remove any excess or tight clothing.

3. If the victim is responsive and not nauseated, provide water or a commercial sports drink (such as Gatorade® or Powerade®).

4. Have the victim lie down and raise his or her legs about 12 in.

5. Cool the victim by applying cool, wet towels to the victim's head and body.

Seek medical care if the condition does not improve within 30 minutes.

Heatstroke

Recognizing Heatstroke

The signs of heatstroke can include the following:

- Extremely hot skin.
- Dry skin (may be wet at first).
- Confusion.
- Seizures.
- Unresponsiveness.

Care for Heatstroke

1. Have the victim stop activity and rest in a cool area.

2. Call 9-1-1.

3. If the victim is unresponsive, open his or her airway, check breathing, and provide appropriate care.

4. Rapidly cool the victim by whatever means possible: cool, wet towels or sheets to the head and body accompanied by fanning, and/or cold packs against the armpits, the sides of the neck, and the groin.

🔌 FIRST AID

Airway Obstruction

Management of Choking—Responsive Victim

1. Check the victim for choking by asking, "Are you choking?" If the victim is unable to respond but nods yes, provide care for the victim.

2. Have someone call 9-1-1.

3. Position yourself behind the victim and locate the victim's navel.

4. Place a fist with the thumb side against the victim's abdomen just above the navel, grasp it with the other hand, and press it into the victim's abdomen with quick inward and upward thrusts. Continue thrusts until the object is removed or the victim becomes unresponsive.

If the victim becomes unresponsive, call 9-1-1 and give CPR. Each time you open the airway to give a breath, look for an object in the mouth or throat and, if seen, remove it.

Adult Cardiopulmonary Resuscitation (CPR)

1. Check responsiveness by tapping the victim and asking, "Are you okay?" If the victim is unresponsive, roll the victim onto his or her back.

2. Have someone call 9-1-1, and have someone else retrieve an AED, if available.

3. Open the airway using the head tilt–chin lift method (lift the chin with one hand and tilt the head back with the other hand).

4. Check for breathing for 5 to 10 seconds by looking for chest rise and fall and listening and feeling for breathing. If the victim is breathing and no spine injuries are suspected, place him or her in the recovery position. If the victim is not breathing, go to the next step.

🔌 FIRST AID

5. Give two rescue breaths (1 second each), making the chest rise. To do this, pinch the victim's nose, place your mouth over the victim's mouth, and give two breaths (1 second each), pausing between each breath. If the first breath does not make the chest rise, re-tilt the head and try the breath again and then proceed to the next step. If both breaths make the chest rise, go to the next step.

6. Perform CPR.

   - Place the heel of one hand on the center of the chest between the nipples. Place the other hand on top of the first hand.

   - Depress the chest 1.5 to 2 in.

   - Give 30 chest compressions at a rate of about 100 per minute, allowing the chest to return to its normal position after each compression.

   - Open the airway, and give two breaths (1 second each).

7. Continue cycles of 30 chest compressions and two breaths until an AED is available, the victim starts to move, EMS takes over, or you are too tired to continue.

Heart Attack

Recognizing a Heart Attack

Prompt medical care at the onset of a heart attack is vital to survival and the quality of recovery. This is sometimes easier said than done because many victims deny they are experiencing something as serious as a heart attack. The signs of a heart attack include the following:

- Chest pressure, squeezing, or pain lasting 2 minutes or longer; it may come and go.

- Pain spreading to either shoulder, the neck, the jaw, or either arm.

- Dizziness, sweating, nausea, or shortness of breath.

⚕ FIRST AID

Care for a Heart Attack

1. Seek medical care by calling 9-1-1. Medications to dissolve a clot are available but must be given early.

2. Help the victim into the most comfortable resting position, usually sitting with the legs up and bent at the knees. Loosen clothing. Be calm and reassuring.

3. If the victim is alert, able to swallow, and not allergic to aspirin, give one adult aspirin or two to four chewable children's aspirin.

4. If the victim has been prescribed medication for heart disease, such as nitroglycerin, help the victim to use it.

5. Monitor the victim's breathing.

Stroke

Recognizing a Stroke

- Sudden weakness or numbness of the face, an arm, or a leg on one side of the body.
- Blurred or decreased vision, especially on one side of the visual field.
- Problems speaking.
- Dizziness or loss of balance.
- Sudden, severe headache.

🔌 FIRST AID

Care for a Stroke

1. Call 9-1-1.
2. If the victim is responsive, lay the victim on his or her back with the head and shoulders slightly elevated.
3. If the victim is unresponsive, open the airway, check breathing, and provide care accordingly. If the unresponsive victim is breathing, place the victim on his or her side to keep the airway clear.

Notes:

Notes:

Notes:

Notes:

The **Gold** Standard in Electrical References

STOCK UP ON THE MOST TRUSTED ELECTRICAL REFERENCE RESOURCES

Ugly's Electrical References, 2020 Edition
978-1-284-19453-1
go.uglys.net/electrical

Ugly's Electric Motors & Controls, 2020 Edition
978-1-284-19455-5
go.uglys.net/motors

Ugly's Conduit Bending, 2020 Edition
978-1-284-20114-7
go.uglys.net/conduit

Ugly's Residential Wiring, 2020 Edition
978-1-284-20484-1
go.uglys.net/residential

Ugly's Electrical Safety and NFPA 70E, 2018 Edition
978-1-284-11940-4
go.uglys.net/safety

For information about bulk pricing or custom sales contact
Sales@uglys.net

www.uglys.net

The **Gold** Standard in Electrical References

IS NOW YOUR GO-TO APP

FEATURING

- Calculators
- Video tutorials for conduit bending
- Complete eBook
- Searchable

- NEMA Wiring configurations
- Works offline
- Regularly updated with new features

Download Ugly's Electrical References today for your Apple or Android device

 www.facebook.com/uglysbooks   www.uglys.net